Series on Technology
and Social Priorities

NATIONAL ACADEMY
OF ENGINEERING

Technology and Global Industry

Companies and Nations in the World Economy

Bruce R. Guile and Harvey Brooks
Editors

NATIONAL ACADEMY PRESS
Washington, D.C. 1987

National Academy Press • 2101 Constitution Avenue, NW • Washington, DC 20418

The National Academy of Engineering was established in 1964, under the charter of the National Academy of Sciences, as a parallel organization of outstanding engineers. It is autonomous in its administration and in the selection of its members, sharing with the National Academy of Sciences the responsibility for advising the federal government. The National Academy of Engineering also sponsors engineering programs aimed at meeting national needs, encourages education and research, and recognizes the superior achievements of engineers. Dr. Robert M. White is president of the National Academy of Engineering.

The president of the National Academy of Engineering is responsible for the decision to publish an NAE manuscript through the National Academy Press. In reviewing publications that include papers signed by individuals, the president considers the competence, accuracy, objectivity, and balance of the work as a whole. In reaching his decision, the president is advised by such reviewers as he deems necessary on any aspect of the material treated in the papers.

Publication of signed work signifies that it is judged a competent and useful contribution worthy of public consideration, but it does not imply endorsement of conclusions or recommendations by the NAE. The interpretations and conclusions in such publications are those of the authors and do not purport to represent the views of the council, officers, or staff of the National Academy of Engineering.

Funds for the National Academy of Engineering's Symposium Series on Technology and Social Priorities are provided by the Andrew W. Mellon Foundation, Carnegie Corporation of New York, and the Academy's Technological Leadership Program. The views expressed in this volume are those of the authors and are not presented as the views of the Mellon Foundation, Carnegie Corporation, or the National Academy of Engineering.

Library of Congress Cataloging-in-Publication Data

Technology and global industry.

(Series on technology and social priorities)
At head of title: National Academy of Engineering.
"Most of the material in this book was presented at a National Academy of Engineering symposium titled 'World Technologies and National Sovereignty' held on February 13 and 14, 1986"—Pref.
Includes index.
1. Technological innovations—Economic aspect—Congresses. 2. Technology and state—Congresses. I. Guile, Bruce R. II. Brooks, Harvey. III. National Academy of Engineering. IV. Series.
HC79.T4T438   1987   338'.06   87-7765
ISBN 0-309-03736-0

First Printing, June 1987
Second Printing, August 1989
Third Printing, November 1990

Preface and Acknowledgments

Many of the strains in today's world lie in the conflict between a global economy that is more and more integrated and a political environment in which national sovereignty is still the dominant motivation. Production and delivery of goods and services is increasingly transnational and will become more so as communications and transportation capabilities increase and their costs continue to decline relative to other costs. Additionally, many new technologies require global markets to recover R&D and initial production costs.

Nations are, however, still reluctant to depend on other nations for key manufacturing inputs. The concept of key technologies that each nation feels it must master inside its own boundaries in order to retain its political independence remains a driving force in international economic relations. Additionally, increasing international economic activity has brought the technological activities of corporations and governments into closer relationships than ever before. National independence is becoming more and more problematic in an interdependent world.

A wide variety of economic, social, and industrial issues are brought forward by the confluence of new technologies, the high-level international interdependence, and the diverse concerns and activities of nations trading in world markets. Modern communications and transportation permit wide dispersal and decentralization of design and production, whereas certain production processes seem to require collocation. How are these two opposing influences reconciled, and how do they vary by industry? What technology policies—by government or by industry—allow the creation of comparative advantage? Does the ability of a nation to handle the adverse social impacts of increased trading in currently nontraded goods lie in the integrity of its social programs or in its skill in international negotiation? This volume addresses these questions and a variety of similar ones.

The principal focus of this volume is on technologies deployed primarily by private firms for commercial purposes—technologies that are altering the structure of the world economy and the location of various types of productive activity—and on the conflicts that arise among national states as a consequence of these shifts. Although the overview and eight chapters in the volume cover a broad range of issues, it is important to mention two issues at the confluence of technology and sovereignty that are not dealt with in this volume. First, there is little explicit treatment of technologies that are inherently global in character, technologies such as satellite communications, remote sensing from space, international commercial air travel, and world oil trade with supertankers. Second, there is no discussion of national security concerns with trade or technology transfer between nations. It is our hope that the focus allowed by not dealing with these two aspects of the global economy yields advantages greater than the disadvantage of not including these obviously important issues.

Most of the material in this book was presented at a National Academy of Engineering symposium titled "World Technologies and National Sovereignty" held on February 13 and 14, 1986. On behalf of the Academy, I would like to thank the advisory committee (listed on page 257) that designed the symposium, and the individuals who participated. I would like to offer special thanks to Robert A. Frosch, John Harris, Cyril Tunis, Jesse H. Ausubel, Ronald S. Paul, Marie-Therese Flaherty, Helen B. Junz, and Pierre R. Aigrain. With regard to the preparation of the manuscript for publication, special thanks are due to H. Dale Langford, the NAE's editor, and to Marjorie D. Pomeroy, administrative assistant in the NAE Program Office.

H. GUYFORD STEVER
Foreign Secretary
National Academy of Engineering

Contents

Technology and Global Industry

Overview

HARVEY BROOKS AND BRUCE R. GUILE

In the last decade it has become increasingly clear that the character of the world economy—and the role of the United States in the world economy—is changing. Two characteristics of global economic change are particularly important. First, over the last 35 years there has been substantial relative growth (in part simply postwar reconstruction) of national economies in Europe and Asia. In the years immediately following World War II the United States dominated world economic affairs, but the U.S. economy is no longer singularly important in the world economy. The United States is now only one element, albeit still a large one, in an increasingly global economy.

The second important change derives, in part, directly from the increasing relative industrialization of other national economies. The growth of other national economies has allowed production and distribution to become increasingly transnational. For a variety of products, fabrication, assembly, distribution, and maintenance activities are organized in such a manner that information, funds, materials, components, final products, and people cross national boundaries as part of everyday commerce. The same is true for service industries, though probably to a lesser degree. Technological advances have played a central role in this economic and technological integration. In particular, technological advance has decreased the relative price of communication and transportation and increased the capacity of transnational systems carrying information, goods, and people.

Increasing global economic and technological integration raises issues

1

concerning the interaction of technological change, economic activity, and the prerogatives of sovereign governments. What are the effects of changing technologies on the production and distribution of goods and services in a global economy? How do technological advances contribute to shifts in the relative competitive advantage of nations, regions, and firms? How do governments and enterprises respond to the dynamic of technological advance in a global economy, and what are the likely consequences, both direct and indirect, of their efforts? This volume explores these and similar questions, focusing primarily on the actions of multinational companies and the policies of industrialized nations.

PATTERNS OF TECHNOLOGICAL CHANGE
AND INDUSTRIAL EVOLUTION

It has long been understood that technological change, through its impact on the economics of production and on the flow of information, is a principal factor determining the structure of industry on a national scale. This has now become true on a global scale. Long-term technological trends and recent advances are reconfiguring the location, ownership, and management of various types of productive activity among countries and regions. The increasing ease with which technical and market knowledge, capital, physical artifacts, and managerial control can be extended around the globe has made possible the integration of economic activity in many widely separated locations. In doing so, technological advance has facilitated the rapid growth of the multinational corporation with subsidiaries in many countries but business strategies determined by headquarters in a single nation.

Fundamental to an understanding of the relation of global industrial structure to technology is the existence of a "technological life cycle." Although the life cycle concept is widely recognized, there is a chronic problem with the unit of analysis. It is not clear whether the appropriate unit of analysis is a highly discrete invention and its subsequent ramifications, a particular product line, or a whole industry. The answer is probably all of the above; particular technological advances may be considered as nested in a larger technological system, which in turn can be viewed as an element of a cluster of related technologies constituting an entire industry.

The chapter by James Utterback introduces the concept of technological life cycles in this volume, a concept that recurs throughout the subsequent chapters. Utterback chooses as his unit of analysis a "productive unit"— something that includes not only a product line or cluster of products closely related by either technology or function but also the processes and procedures used in their production. He examines parallel paths of evolution of technologies and organizations (in several industries) and makes a case for considering industrial structure and effective strategy as directly related to location of both the product and process technology in a technological life cycle. The general issues raised by Utterback are extended to the mature phases of product and process cycles in the chapter by Alvin P. Lehnerd, who provides a dramatic account of how fundamental reconsideration of both product and process design for a mature product line can revolutionize the competitive position of that line in the international marketplace.

Lehnerd describes a program at the Black & Decker Company during the 1970s. The program substantially improved the company's productivity in the manufacture of power tools by designing products for production using new materials and new manufacturing techniques, greatly reducing the number of parts and standardizing components common to the various products within the product line. During the 1970s when non-U.S. manufacturers began to dominate in many traditional manufactured goods, Black & Decker picked a strategy—linked to new technology—that allowed them to become a high-value, low-cost producer in some lines of small power tools. The success of Black & Decker's program raises the possibility that production of many "mature" products could be revolutionized by attention to design for improved manufacturability. Lehnerd's example also suggests that a heavy investment in in-place facilities for manufacturing a mature product becomes a barrier both to product innovation and to the introduction of more advanced low-cost production processes, especially when the mature product line appears to be still doing well in the market. In the case described by Lehnerd, this was exactly the time when a radical redesign of an existing product line and its associated manufacturing process proved essential to the competitive position of the firm.

The relationships between technological advance and industrial structure that Utterback and Lehnerd address are extended to questions of global organization and technology in the chapters by David J. Teece, Yves Doz, and James Brian Quinn.

TECHNOLOGY AND THE STRUCTURE OF GLOBAL INDUSTRY

David Teece's chapter deals with returns to innovation and the arrangements—integration, partnering, and licensing—that determine whether the potential economic returns from an innovation will be realized by the innovator or an imitator. In his discussion, Teece draws on a different set of examples and reinforces and elaborates the points made by Utterback concerning technological trajectories, efficient industry structure, and the importance of matching structure and strategy to the location of each product line in the technological life cycle. Teece also cautions against economywide generalizations about innovation or technology-driven markets. In a world of many nations of nearly equal levels of industrialization and technological prowess, the likely impacts of movement along a technological trajectory are not obvious. For example, the product life cycle has signficantly different consequences for the European automobile industries than for their Japanese and American counterparts. The European market is more of a "niche" market than the Japanese and American markets, which are more commodity-like, and technology plays a different role in the two cases. In Teece's words:

> [T]he product life cycle in international trade will play itself out differently in different industries and markets, in part according to appropriability regimes (the degree of appropriability of the potential economic gains from technological innovation) and the nature of the assets needed to convert a technological success into a commercial one. . . . [I]t is not so much the structure of markets as the structure of firms, particularly the scope of their boundaries, coupled with national policies for the development of complementary assets, that determines the distribution of profits among innovators and imitator-followers [Teece, this volume, p. 94].

Teece also suggests that lack of attention in the United States to aggressive investment in new manufacturing technology may have enabled skillful imitator-followers, particularly in Japan, to appropriate a disproportionate share of the gains from U.S. inventions arising out of its uniquely broad-based R&D programs, both public and private. As reflected also in the chapter by Doz (and in the chapter by Henry Ergas later in the volume), the too-exclusive emphasis of public policy in both Europe and the United States on R&D and technological prowess as the principal remedies for lagging competitiveness may be misplaced; without appropriate complementary assets and capacities that allow a nation to capture returns from innovation, national technological superiority is of little economic consequence.

The trend during the last 30 years has been toward global homogenization of markets and transnational integration of production. Yet there are signs of the emergence of countervailing pressures resulting from technological, managerial, and political developments that appear to be giving a competitive advantage to more localized production and distribution. For example, the relative importance of close interaction between producers and users seems to be growing as products become more complex and customized. Effective product design, prompt maintenance service, and consultation services to customers in increasingly sophisticated applications all require close links between sellers and customers.

Additionally, the increased use of just-in-time inventory principles in production and distribution has enhanced the competitive advantage of collocation of suppliers and distributors with manufacturing operations. The chapter by Yves Doz reviews these countervailing influences in the global economy.

In his analysis Doz points out that trends in new technology permit enormous flexibility and diversity in the way global enterprises are actually managed. Firms with widely dispersed facilities tightly integrated in a global strategy exist side by side with firms having single production sites geared to serving dispersed markets or specialized market niches that are spread globally. Because modern technologies are both flexible and diverse, other factors may be more important than technology as a determinant of organizational structure. The costs associated with production and distribution may be less important in determining the structure of an industry than the organizational and political imperatives of partnership opportunities, investment for market access, or access to localized concentrations of specific technical skills. Therefore, despite forces that push for either fragmentation or homogenization of markets, the dominant characteristic of the structure of industry in the global marketplace may be diversity.

In his chapter on technology and the service industries, James Brian Quinn addresses many of the issues raised by Doz and Teece, but from a somewhat different perspective. His focus is less on the international structure of industries and more on the interaction of technological advance with the evolution of organizational structure in a variety of service industries. In particular, Quinn approaches the delivery of goods and services to the consumer as a long chain of labor, capital, location, and organization, each adding value to create the final product. He uses this framework to challenge the common perception that service industries have both low labor productivity and low productivity growth, add little value, and provide only low-wage, insecure jobs. Quinn offers many examples of the importance of technology to the development and restructuring of service industries, to the emergence of whole new services, and to the impressive growth in productivity in some service sectors.

Both Doz and Teece treat major service activities—finance, transportation, communication, and wholesale and retail activities—as integrated parts of an organization delivering a product to a consumer. Neither distinguishes between value added to a product through the act of assembly (manufacturing) and value added through the act of delivery (service). Technological advance is important in both activities, and in both it can increase efficiency and provide new opportunities for organization. Indeed, the similarities between technological and organizational issues in manufacturing and service industries make distinctions between services and manufacturing seem arbitrary. Quinn argues persuasively that much conventional wisdom about service industries is based on little empirical evidence and a lack of recognition of the heterogeneity of the activities grouped under the heading of services. A reexamination of the sector is at present inhibited by poor statistics and especially by inadequate and obsolete categorization.

The chapter by Quinn also raises the question of whether national success in services can replace manufacturing as the engine of national economic progress and international competitiveness, much as manufacturing had replaced agriculture and resource industries as a source of growth and employment in an earlier period. As illustrated in Quinn's chapter, many service industries—medical care, transportation, communications, and banking, for example—are technologically dynamic and crucial to national economic performance. The services sector now accounts for almost 70 percent of U.S. gross national product, and the United States has persistently enjoyed a positive net balance of trade in services and income from foreign investment. On the other hand, U.S. trade in services, though probably underestimated at present, is small in relation to U.S. trade in manufactures and agricultural products, and it seems unlikely that an industrialized nation the size of the United States can export enough services to cover the cost of importing a predominance of the manufactured goods it demands. Additionally (as discussed in Raymond Vernon's chapter in this volume), the fraction of U.S. GNP accounted for by manufacturing output, when proper allowance is made for the lower rate of price increase for manufactured than for nonmanufactured output, has remained approximately constant—between 20 and 23 percent with no clear trend up or down—for the past 30 years (*Economic Report of the President 1987*, Table B-11, p. 257). This observation belies the argument that the United States is rapidly becoming solely a service producer. Indeed, the degree to which the domestic economy of any large industrialized nation can become a "service economy" is further complicated by the complex technological and economic interdependence of services and manufacturing.

Service industries are both suppliers to and buyers from manufacturing industries. As buyers of manufactured goods, service industries are increasingly dependent on the rapid deployment of technology-intensive capital goods for improving productivity. It is not clear whether any nation can remain competitive in services if it becomes too dependent on foreign sources for this complementary capital embodying the most advanced technology. There is, of course, the potential that national governments may use various kinds of controls on the export of services-related capital goods to ensure the competitive advantage of domestically based services in international trade. On the other hand, the developers of much of the technology used in services are manufacturers who are strongly motivated to introduce and sell their technologies as widely and rapidly as possible worldwide to recover heavy development costs. Proximity to the sources of innovation in services-related capital goods may or may not contribute to national competitive advantage in any service industry.

As suppliers to manufacturing industries, many service activities may have to be intimately linked to customers and adapted to unique local needs if they are to be effective. If that is indeed the case, then a vital manufacturing sector may be a necessary prerequisite for service industry development and hence for national economic health. Manufacturing competitiveness may in its turn be critically dependent on the efficiency and cost of the services locally available to manufacturing plants—services essential to the smooth functioning of tightly integrated manufacturing and distribution systems. For example, the operation of manufacturing and distribution systems with minimal inventory costs and buffer stocks requires a highly efficient low-cost service infrastructure.

It is not clear for which sectors the complicated linkages described above are most important and which goods or services, if any, a large nation can afford to import over the long run from distant locations. Still unanswered, therefore, is whether the "postindustrial society," the "information society," or the "service economy" are catch phrases that rationalize the relative decline of manufacturing employment, or whether they truly represent the "wave of the future" and a sufficient foundation of future national prosperity and wealth. What is clear is that the application of technological advance to service industries can be central to improved economic performance in both service and manufacturing industries. To keep pace with productivity improvement in other industrialized nations, a nation must direct its trade and economic policies toward supporting fast and flexible deployment of technologies in service industries regardless of the location of the source of the technology.

Taken together, the chapters by Teece, Doz, and Quinn do not suggest a trend toward either homogenization or segmentation of world markets and world industries. While some product and service markets are becoming global, driven by ever-increasing economies of scale, other markets are fragmenting and differentiating. New manufacturing and service delivery technologies, new methods of work organization, and a new importance of local market responsiveness all can decrease the significance of scale economies and favor decentralized production. The long-term norm may be loose global coordination and frequent temporary alliances among particular units in different countries for different, and usually highly product-specific and market-specific, purposes. The global economy appears to be moving toward a complex (and often highly interdependent) coexistence of centralized and decentralized markets and production systems.

Corporations—in their national or transnational activities—depend on the laws, infrastructure, and political stability provided by national governments. In turn, governments of industrialized nations in the noncommunist world depend on private enterprise to provide employment, income growth, and goods and services for the nation's citizenry. The growth of transnational organization in production raises new concerns about the interdependence of nations and companies.

NATIONAL ECONOMIC DEVELOPMENT AND MARKET-DRIVEN DEPLOYMENT OF WORLD-SCALE TECHNOLOGIES AND INDUSTRIES

The growth of industrialized economies in Europe and Asia since World War II has eroded the importance to the world economy of both U.S. domestic economic policies and unilateral U.S. foreign investment and trade policies. This change has consequences for virtually every aspect of the world economy as the importance of multilateral negotiation and agreement grows apace. Though national foreign policies have a variety of purposes, it is almost always true that the primary goal of national participation in international economic affairs is national economic development. Recently, the concerns of industrialized nations over economic development in a world economy have been expressed mostly in terms of national competitiveness. However, as economic institutions become more global in scope, whether through networks of alliances across national boundaries or through large centrally controlled transnational corporations, the concept of a competitive national economy becomes uncertain and obscure.

One measure of competitiveness may be the average level of real wages that labor can command in a given country (and the potential for future growth of this level), but it can be argued that measures such as employment growth, technological capability, productivity growth, or corporate profitability are better proxies for what is meant by competitiveness. These measures do not, however, reflect the same concept of competitiveness. Though real wages are a gross indicator of standard of living, employment growth may be a better measure of the opportunities available to the citizenry. Technological capability and productivity growth relate to the productive resources physically located in a given country's territory, whereas corporate profitability reflects the performance of firms with their headquarters and primary ownership in a given country, regardless of the location of production or distribution.* These various measures of economic development—and the policy goals implicit in the measures—are central to national policy debates. Three chapters in this volume—by Raymond Vernon, Henry Ergas, and Lewis M. Branscomb—deal extensively with national economic development policies.

Vernon's chapter reviews some basic and inevitable changes in the position of the U.S. economy in the global marketplace and addresses many of the concerns expressed by the U.S. public and U.S. policymakers about international competitiveness. Vernon raises two issues that are not discussed elsewhere in the volume. The first is a concern with the internal distribution of the national costs associated with U.S. participation in an open global economy. Particularly important is his assessment of trade-offs within the U.S. economy. Some industries and groups in the United States have suffered from the exposure of U.S. markets to foreign producers, such as those associated with the steel and auto industries; but others have benefited, such as low-income groups—who enjoy better prices for clothing, household appliances, food, and other basic goods—or workers in successful export industries. By the same token, factory workers in some traditional industries may have fared poorly, but management consultants and computer software specialists have done well.

The second policy issue unique to Vernon's chapter in this volume is his assessment of the challenge to U.S. policymakers to avoid triggering a cascading sequence of beggar-thy-neighbor actions that would change the policies of governments from the fostering of positive-sum games to mutually destructive actions designed to protect the interests of politically influential domestic constituencies. The U.S. political tendency is to respond to localized domestic industrial distress with political action. The current furor over the U.S. trade deficit is a good example. As the U.S. trade balance has worsened, there has arisen a widespread belief in the United States that other nations are not playing by the rules of the open

---

*Because of the ambiguity of the term "competitiveness," the picture with regard to U.S. competitiveness is not clear. By a number of these measures—in particular productivity growth and increases in real wages—the U.S. economy has not been performing as well as other industrial economies in the last 15 years (Scott, 1985; President's Commission on Industrial Competitiveness, 1985). In employment growth, however, the U.S. economy has done better than other industrial economies, having created many more new jobs over the same period, though questions have been raised about the "quality" of these jobs (Bluestone and Harrison, 1986). In scientific and technological capability, the United States is still the world leader (Brooks, 1986), but there are significant questions (as raised by Teece in this volume) regarding U.S. application of technology. Finally, there are analyses that indicate that, although the United States as a location of production may have lost world market share, U.S.-based multinationals have gained market share in world markets (Lipsey and Kravis, 1985).

trading system that the United States helped to establish after World War II. The perception that the United States respects these rules while other nations do not has generated a chorus of demands for unilateral or coercive U.S. actions to create a "level playing field." Are these demands legitimate? Most scholarly studies of nontariff trade barriers indicate that the fraction of the total value of U.S. imports affected by such barriers is as large as or greater than the fraction of imports affected in notable rivals such as Germany and Japan (Saxonhouse, 1983; Cline, 1984). If the United States is different from its industrialized competitors in this regard, it is only in the fact that the barriers appear to follow a less coherent or consistent pattern than those of some other industrialized nations.

The concern in the United States is not an unusual or surprising response to trade problems. There is a tendency for every nation to see itself as unfairly disadvantaged by world competition in sectors in which it is doing poorly while taking for granted its success in sectors where it is doing well. Thus, each nation attempts to intervene politically in these disadvantaged areas and is troubled by the inadequacy of its political influence over the policies and actions of other nations.

Among the policies that nations use for economic development are those to promote technological advance. The chapter by Henry Ergas presents a cross-national comparison of technology policies. The thrust of Ergas's argument is that various strategies are open to countries, or to businesses within countries, based on where they choose to seek competitive advantage in the product or technology cycle. In Ergas's assessment, the United States, France, and the United Kingdom have chosen (if such an active word can be applied to a set of policies that have evolved in a fairly decentralized manner) to seek competitive advantage in the stage at which a new technology is just emerging, whereas Germany, Switzerland, and Sweden have chosen to configure their public policies and their industrial structure to take maximum advantage of the more mature phases of development in products and processes. Japan, Ergas argues, has chosen to try to profit from the "consolidation" or "take-off" phase, and in large measure its strategy is something of a hybrid between the emerging technology strategy of the United States and the diffusion strategy of Germany.

Ergas writes:

> This discussion suggests that there are different paths to happiness, as countries' institutional structures and social arrangements facilitate specialization in differing stages of technological evolution. Each of these stages has advantages and disadvantages in providing for the growth of real income, but countries also differ in the extent to which they succeed in securing the greatest benefits from any given pattern of specialization.
>
> [L]ocation on a technological trajectory may be less important than the efficiency

with which the advantages of that location are pursued. This, in turn, depends on institutional features (broadly defined) that may be more or less appropriate for a given pattern of specialization [Ergas, this volume, pp. 230-231].

This categorization is important from the U.S. perspective, especially when combined with Ergas's argument that the emerging-technology phase does not usually produce large gains in per capita income or value-added per worker. The implication is that there may be, from an economic development perspective, a comparative overemphasis in the United States on creativity, originality, novelty, and sophistication at the leading edge of technological advance. This overemphasis comes at the expense of what could be called the "creative imitation" or rapid incremental improvements that the Japanese are especially good at.

It is worth noting that Vernon and Ergas, writing from macroeconomic and public policy perspectives, both reinforce points made by Teece from a microeconomic perspective. In a world in which technological innovators cannot hope to capture more than a small fraction of the gains from their innovations, and in which the successful exploitation of a technological advance depends on tapping global markets, a national economy that invests in the creation of new technologies must constantly ask itself where the economic returns to such advances are likely to be captured. The ability of a nation to generate technological advances is insufficient by itself, and may not even be essential, for improving national competitive position.

Branscomb, in the final chapter in the volume, compares the technology development and deployment strategies of companies and governments. Branscomb discusses the existence of both synergy and conflict in the interests of nations and corporations. Governments and transnational companies share an important common interest in economic growth and development, but each has ancillary goals not necessarily consistent with those of the other. Governments care—for a variety of legitimate reasons—about national self-sufficiency, whereas corporations care primarily about profitability and autonomy of action. Though these interests do not always conflict, there are inevitably situations where the goals and methods of the two types of organizations diverge. In other words, the imperatives of global economic and technological interdependence—often manifest in transnational production and distribution activities—sometimes run against legitimate nationalistic concerns.

In a global economy, the autonomy and importance of multinational corporations can restrict the ability of national governments to carry out independent economic and social policies within their boundaries. Therefore, an ongoing important international policy question is: What mechanisms will allow a group of nominally independent sovereign nation-states—working with a parallel group of nominally autonomous transnational companies—to deal with a global-scale economy in a way that is just and equitable for all the different publics involved? There are no simple or even obvious solutions. National limitations on the autonomy of multinationals may come at the expense of national and global economic growth. On the other hand, corporate autonomy may come at the expense of painful domestic adjustments and localized welfare losses as well as losses of national self-esteem and cultural autonomy. The challenge is to develop an international political regime that provides for negotiation over the needs of different national constituencies without choking off the open exchange of goods and services. This has important implications for the policies of nations and companies in relation to political constituencies; a stable system of governance for the international economy cannot long accommodate to severe adverse economic effects on individual nations or influential constituencies within nations. In particular, effective social policy—temporary assistance to disadvantaged groups, for example—may help to mitigate constituency resistance to change and afford national negotiators a freer hand in representing truly national rather than vocal parochial interests.

CONCLUSION

Taken together, the chapters in this volume raise many issues about patterns of technological change and evolution in the structure of organizations—and styles of management—in the global economy. These chapters contribute to a growing literature—built on ideas expressed by N. D. Kondratiev and Joseph Schumpeter in the first half of this century—that explicitly links technological trajectories or life cycles to industrial development. Because of the complexity created by nested and overlapping technological advances, the interpretation of what constitutes a technological trajectory is rather vague and, though there have been substantial contributions to our understanding (Abernathy and Utterback, 1978, and Dosi, 1982, for example), no composite theory has ever been worked out in detail. Despite this, there appears to be some common understanding of a three-stage pattern of technological development. The *first stage*— the "emergent" or "fluid" phase—is viewed as a period of great ferment during which the various actors, particularly inventors and users, carry out a trial-and-error search for the application of an initial concept that works—both technically and in terms of customer acceptance. In this phase there are often many competing firms and technical ideas and no clearly superior design.

The *second stage* is characterized by the emergence of a dominant design (or application that appears to meet the requirements of the marketplace). At this point the pace of diffusion of the new technology quickens, and at first many new competitors "swarm" into the market. As diffusion continues, price competition becomes more important and there is less product differentiation on the basis of product characteristics. The pure economics of production and delivery come to dominate competition. Simultaneously product innovation becomes more incremental, based on the now dominant design concept, and there is more stress on innovation to bring down costs and increase quality and uniformity of the product. The search for improvements narrows, but the rate of reduction of costs accelerates and, with it, the rate of market penetration because of price elasticity of demand. At the same time the race for cost-reducing improvements drives many competitors out of the market, and a much more stable division of market shares among the remaining competitors results.

The *third stage* is reached when the market begins to saturate, new applications and new markets give way to replacement of previous generations of the same technology, and further cost-reducing improvements become harder and more expensive to find. What happens, or should happen, in this mature phase of a life cycle is the subject of much less agreement. It is a period in which the leading competitors are much more vulnerable to the appearance of a radical innovation, which may constitute the initiation of a new technological paradigm. In this phase the organization tends to be optimized for mass producing and marketing a commodity-like product. This form of organization is likely to be unsuitable for introducing and rapidly improving a product or a new manufacturing process that is in its dynamic growth phase. Lehnerd's example from the Black & Decker Company seems to be the exception rather than the rule.

Although, as mentioned above, there is little agreement on the specific characteristics of the technological life cycle and the level of aggregation of economic activity to which it is relevant, the loosely defined notions of a technological trajectory or product life cycle have proved useful in dividing technological advance into stages that can be linked to trade patterns, economic structure, and national technological strategies in the global economy. The product life cycle theory developed by Raymond Vernon in the late 1960s (Vernon, 1966), and subsequently elaborated by many authors, is a prime example. The chapters in this volume are consistent with that tradition. They strongly suggest that the technological character of product lines, production processes, and delivery systems in an industry evolves in a consistent, though subtle, manner in a way that dramatically influences both the range of viable business strategies and the likely market outcomes in the global economy.

In addition to addressing industrial and technological change, the chapters in this volume delineate a chronic tension in global economic and technological affairs. The principles of relatively unrestricted world trade—carried out most often through multinational firms and benefiting consumers and in many cases the firms' managers and shareholders—conflict with the legitimate interests of important producer and other interest groups within nations. With increasing globalization of economic activity, bilateral and multilateral negotiations over trade and foreign investment practices—already an important aspect of national foreign policy—will be increasingly important components of national domestic economic and social policy. How can a national government accommodate the interests of important groups that are seriously affected by developments in the international economy? Are policies targeted toward particular key industries and technologies more significant for national economic health than government support for education and basic research, the development of generic technologies, or the upgrading of basic infrastructure? Is a multinationally coordinated approach to more general policies such as macroeconomic policies, national tax structures, regulatory philosophies, policies toward human resource development, and labor market adjustment a desirable goal? These questions will never be settled in the large; future policy will exist primarily in the resolution, or lack of resolution, on specific negotiations. The questions, however, are likely to be important national policy issues for decades.

REFERENCES

Abernathy, W. J., and J. M. Utterback. 1978. Patterns of industrial innovation. Technology Review 80:7(June-July):40-47.

Bluestone, B., and B. Harrison. 1986. The Great American Job Machine: The Proliferation of Low Wage Employment in the U.S. Economy. A study prepared for the Joint Economic Committee, December.

Brooks, H. 1986. National science policy and technological innovation. Pp. 119-167 in The Positive Sum Strategy, R. Landau and N. Rosenberg, eds. Washington, D.C.: National Academy Press.

Cline, W. R. 1984. Exports of Manufactures from Developing Countries. Washington, D.C.: Brookings Institution.

Dosi, G. 1982. Technological paradigms and technological trajectories. Research Policy 11(3):147-162.

Economic Report of the President, 1987. Washington, D.C.: U.S. Government Printing Office.

Lipsey, R. E., and I. B. Kravis. 1985. The Competitive Position of U.S. Manufacturing Firms. National Bureau of Economic Research Working Paper 1557. Cambridge, Mass.: National Bureau of Economic Research.

President's Commission on Industrial Competitiveness. 1985. Global Competition: The New Reality. Volumes 1 and 2. Washington, D.C.

Saxonhouse, G. 1983. The micro- and macroeconomics of foreign sales to Japan. Pp. 259-263 in Trade Policy in the 1980s, W. R. Cline, ed. Cambridge, Mass.: MIT Press.

Scott, B. R. 1985. U.S. competitiveness: Concepts, performance, and implications. Pp. 13-69 in U.S. Competitiveness in the World Economy, B. R. Scott and G. C. Lodge, eds. Boston, Mass.: Harvard Business School Press.

Vernon, R. 1966. International investment and international trade in the product cycle. Quarterly Journal of Economics 80(2):190-207.

Innovation and Industrial Evolution in Manufacturing Industries

JAMES M. UTTERBACK

Historically, studies of innovation have had a linear viewpoint. That is, they have seen innovation as something that begins with a company possessing a certain technology and then investing in that technology, and the accompanying ideas, and implementing them in the market. This approach, however, assumes that all innovations occur in the same way in all companies and disregards the fact that organizations change throughout their lifetimes. It also fails to distinguish between product and process innovations, each of which may follow a different path. In short, the interaction of technological change and the marketplace is much more complex and dynamic than linear models can describe. The dynamic model discussed below describes how change in product innovation, process innovation, and organizational structure occurs in patterns that are observable across industries and sectors. The dynamic model allows consideration of the different conditions required for rapid innovation and for high levels of output and productivity. The argument describing this model is built on historical studies of innovations in their organizational, technical, and economic settings. Such data are necessarily incomplete, but at the same time, they yield a rich variety of insights.

Parts of this chapter draw upon the following previously published sources: Abernathy and Utterback, 1978; Hill and Utterback, 1979; Utterback, 1978; and Utterback and Abernathy, 1975.

16

UNIT OF ANALYSIS

Product and process innovation are inextricably interdependent; in considering manufacturing innovation, both a product line and an associated production process must be taken together as the unit of analysis. Termed a productive unit in this chapter, this unit of analysis is a slightly different concept than a business or a firm. For a simple firm or a single-product firm, the productive unit and the firm would be one and the same. In a diversified firm, however, a productive unit would usually report to a single operating manager and normally be a separate operating division. When the word "firm" is used in this chapter, it should be taken narrowly to mean productive unit as defined here.

Competition in the marketplace is not only between firms, but often between products or product lines. Even an enterprise classified as a single industry might find itself competing with many disparate groups of firms with different product lines or lines of business. Thus, to group productive units sensibly into industry or market segments, one must ask: In what product lines do units view each other as direct competitors? Within a segment, productive units that view each other as direct competitors face a similar business environment and set of competitive requirements for their technology. The terms "industry" and "market segment" will be used here in this limited sense.

A key idea is that productive units may be arranged in a dependent hierarchy from final market to equipment and materials suppliers. Thus, what is viewed as a product innovation by a unit at one level is part of the production process or product of a unit at the next higher level (Abernathy and Townsend, 1975). This means that most innovations affect productivity directly. It also means that the markets to which innovations respond are often defined by the characteristics of other firms' production processes. Operations management and management of technological innovation and change are inextricably linked.

The fact that one firm's product is another's manufacturing equipment or material, and the fact the major product changes are often introduced from outside an established industry and viewed as disruptive by the existing competitors, means that the standard units of analysis of industry— firm and product type—are of little use, for as technology changes, the meaning of these terms also changes. Analysis of change in the textile industry requires that productive units in the chemical, plastics, paper, and equipment industries be included. Analysis of electronics firms requires review of the changing role of component, circuit, and software producers as they become more crucial to change in the final assembled product. Major change at one level works its way up and down the chain because of the interdependence of product and process change within and among productive units. Knowledge of the production process as a system of linked productive units is a prerequisite to understanding innovation and competition in an industrial context.

Earlier work on the management of technology has focused at a micro level, dealing with similarities among particular successful cases of product or process innovation (Utterback, 1975), whereas work on the economics of technological change has focused at a macro level, dealing with changes in productivity and technology among industries (Rosenbloom, 1974). Neither has aimed at understanding the dependence of product innovation on process innovation and its crucial importance for operations management and strategy. Use of the idea of a productive unit as the unit of analysis requires focusing on their critical interaction, both within a unit and between units linked by physical flows of equipment, material, and parts (Abernathy and Townsend, 1975).

PRODUCT INNOVATIONS

What is needed is a view of innovation that will aid the decision-making process of company managers, government policymakers, and researchers. Out of this need has arisen a theory holding that the interaction between technology and the marketplace is much more complex and dynamic than the linear view would have us believe. It is our contention here that the conditions required for rapid innovation are extremely different from those required for high levels of output and productivity: Under demands for rapid innovation, organizational structure will be fluid and flexible, whereas under demands for high levels of output and productivity, organizational structure will be standardized and inflexible. Thus, a firm's innovation attempts will vary according to its competitive environment and its corresponding growth strategy. It will also be affected by the state of development of both its production technology and that of its competitors (Abernathy and Utterback, 1978). Therefore, we can expect to see different creative responses from productive units facing different competitive and technological challenges, which, in turn, suggests a change in the way of viewing and analyzing possible policy options for encouraging innovation.

A dynamic model of innovation (Figure 1) includes a pattern of sequential and cross-sectoral change in product innovation, process innovation, and organizational structure. Firms that are new to a product area will exhibit a fluid pattern of innovation and structure. As the market develops, a transitional pattern will emerge. Finally, the market stabilizes, fostering a specific pattern of behavior. Therefore, a radical innovation— one that can create new businesses and transform or destroy existing ones—

is often the result of the addition of entirely new requirements to a previously stable set of dimensions (Normann, 1971).

In the fluid phase of a firm's evolution, the rate of product change is expected to be rapid, and operating profit margins are expected to be large. The few existing competitors will be either small new firms or older firms entering a new market based on their existing technological strengths. A firm might be expected to emphasize unique products and product performance in anticipation that the new capability will expand customer requirements. The new product technology will often be crude, expensive, and unreliable but will fill a function in a way that is highly desirable in some market niche. Prices and profit margins per unit will be high, because the product often has great value in a user's application.

Several studies have shown that the performance criteria that serve as a primary basis for competition change from ill defined and uncertain to well articulated as a firm travels through the various states of development (Frischmuth and Allen, 1969). In emerging product areas, there is a proliferation of product performance dimensions. These frequently cannot be stated quantitatively, and even the relative importance or ranking of the various dimensions may be unstable. Thus, because most product innovations will be market-stimulated, there will be a high degree of uncertainty about their potential. This can be called target uncertainty.

Although the total amount of research and development (R&D) in a sector may be large, its focus will be diffuse. This is called technical uncertainty. The expected value of any R&D investment is reduced by the combined effect of target uncertainty and technical uncertainty.

Technology to meet needs will come from many sources, including customers, consultants, and other informal contacts, because fluid units tend to rely heavily on diverse, external sources of information. However, the critical input will not be state-of-the-art technology but new insights about needs (von Hippel, 1977); innovations will originate in units with intimate knowledge of users and user needs.

As both producers and users of a product gain experience, target uncertainty lessens and product innovation enters the transitional state. The usefulness of the new product is increasingly better understood, and it may take on a variety of new forms to serve other parts of the market. Additional improvements and innovations incorporating new components and systems concepts may be required to expand its possible uses and sales. A greater degree of competition based on product differentiation usually develops, and dominant product designs may begin to emerge.

At the same time, forces that reduce the rate of product change and innovation are beginning to build up. As obvious improvements are introduced, it becomes increasingly difficult to better past performance, users

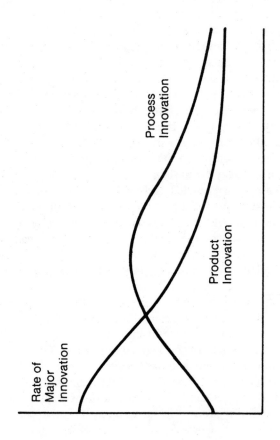

	Fluid Pattern	Transitional Pattern	Specific Pattern
Competitive emphasis on	Functional product performance	Product variation	Cost reduction
Innovation stimulated by	Information on users' needs and users' technical inputs	Opportunities created by expanding internal technical capability	Pressure to reduce cost and improve quality
Predominant type of innovation	Frequent major changes in products	Major process changes required by rising volume	Incremental for product and process, with cumulative improvement in productivity and quality
Product line	Diverse, often including custom designs	Includes at least one product design stable enough to have significant production volume	Mostly undifferentiated standard products
Production processes	Flexible and inefficient; major changes easily accommodated	Becoming more rigid, with changes occurring in major steps	Efficient, capital-intensive, and rigid; cost of change is high
Equipment	General-purpose, requiring highly skilled labor	Some subprocesses automated, creating "islands of automation"	Special-purpose, mostly automatic with labor tasks mainly monitoring and control
Materials	Inputs are limited to generally available materials	Specialized materials may be demanded from some suppliers	Specialized materials will be demanded; if not available, vertical integration will be extensive
Plant	Small-scale, located near user or source of technology	General-purpose with specialized sections	Large-scale, highly specific to particular products
Organizational control is	Informal and entrepreneurial	Through liaison relationships, project and task groups	Through emphasis on structure, goals, and rules

FIGURE 1 A dynamic model of innovation. Reprinted with permission from Abernathy and Utterback (1978), *Technology Review*, copyright 1978.

develop loyalties and preferences, and the practicalities of marketing, distribution, maintenance, advertising, and so forth demand greater standardization. Innovations leading to better product performance become less likely unless the improvement is easy for the customer to evaluate and compare, for firms will attempt to maximize their sales and market share by defining their needs based on those of the customer.

The reduction in target uncertainty that comes from greater diffusion of product use allows a correspondingly greater degree of technical uncertainty to be tolerated. Therefore, larger R&D investments will be justified—for advanced technology will become a major source of further product innovation. At some point, before the cost of technological innovation becomes prohibitively high, and before increasing cost competition erodes margins below levels that can support large categories of indirect expense, the benefits of R&D probably reach a maximum.

The emergence of a dominant product design that enforces standardization marks the beginning of the specific state. Such product design milestones can be identified in many product lines; sealed refrigeration units for home refrigerators and freezers, effective can-sealing technology in the food canning industry, and the standardized diesel locomotives in the locomotive and railroad industry are but a few examples.

George White (1978) contends that dominant designs can be recognized in the early stages of their development. He suggests that dominant designs will usually display several of the following qualities:

- Technologies that lift fundamental technical constraints on the art without imposing stringent new constraints.
- Designs that enhance the value of potential innovations in other elements of a product or process.
- Products that ensure expansion into new markets.
- Products that build on existing operations rather than replacing them.

The dominant new product design signals a significant transformation, affecting the type of innovation that follows it, the source of information, and the size, scope, and use of formal research. As the productive unit evolves into this specific state, the set of competitors often becomes an oligopoly and competition begins to shift to product price, which means that product design and process design become more and more closely interdependent as a line of business develops. Margins are reduced, and production efficiency and economies of scale become emphasized. Consequently, the requirements for the market become simpler and more precise. As price competition increases, production processes become more capital-intensive and may be relocated to achieve lower costs. This relocation may even shift capacity overseas (Vernon, 1966; Wells, 1972).

Because investment in existing process equipment is high, and product and process change are interdependent, both product and process innovations in the specific state are usually incremental. Under these conditions, however, both product and process features are well articulated and easily analyzed, and the conditions necessary for the application of scientific results and systems techniques are present. Unfortunately, the payoff required to justify the cost of change is large, whereas potential benefits are often marginal; innovations typically will be developed by equipment suppliers for whom the incentives are greater and adopted by the larger user firms (Abernathy, 1976; Abernathy and Wayne, 1974). Thus, as the product market shifts from fluid to transitional to specific, the locus of major product innovation may shift from user to manufacturer to equipment supplier (see Figure 2).

PROCESS INNOVATIONS

A production process is the system of process equipment, work force, task specifications, material inputs, work and information flows, and so forth employed by a unit to produce a product or service. In the fluid state, the productive unit will typically be small, with limited resources. Order backlogs may rise rapidly, even though the market is small, reflecting the unit's limited capacity. The novelty of the product may mean that the unit will be the sole supplier for a limited period of time.

In this situation, the unit will attempt to expand rapidly in the simplest way possible. The emphasis will be on highly skilled and flexible labor, and the process itself will be composed largely of unstandardized and manual operations, or operations that rely on general-purpose equipment. The adaptations made to equipment by the firm will be minor, as in a job shop, and the problems of coordination and control will be similar. Capacity levels will be poorly defined. Such a system necessarily is inefficient. Greater volume will be achieved through paralleling existing processes and improving manual operations. There will be few scale barriers to entry into the business.

As a small purchaser, the unit will usually have little influence over its suppliers. Raw materials and parts will be used as available; if new materials or parts are produced for the unit, their quality may vary widely. Variations in input quality and product design are compensated by the considerable flexibility in the types of tasks each individual and piece of equipment can perform.

When significant-enough volume is achieved in one or more product lines to encourage standardization, the production process enters the transitional state. Major process change then occurs at a rapid rate. Production

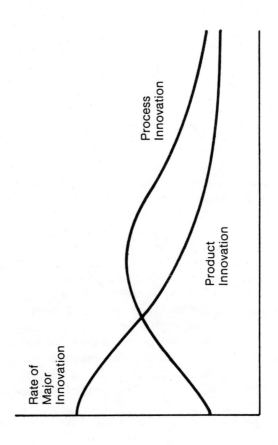

	Fluid Pattern	Transitional Pattern	Specific Pattern
Competitive emphasis on	Functional product performance	Production variation	Cost reduction
Innovation stimulated by	Information on users' needs and users' technical inputs	Opportunities created by expanding internal technical capability	Pressure to reduce cost and improve quality
Product line	Diverse, often including custom designs	Includes at least one product design stable enough to have significant production volume	Mostly undifferentiated standard products
Performance criteria are	Ill-defined and uncertain targets for innovation; many often qualitative criteria	Becoming stable with each product occupying a somewhat different position vis-à-vis criteria	Well-defined and monotonic; cost- and quality-related dimensions predominate
Uncertainty concerns	The relevance of outcomes that might be achieved; demands are ill defined	The balance of market and technical factors as appropriate targets for R&D are clearer	Primarily technical issues as market demands are well known
Source of technology is	Often a user of the product	Often a manufacturer of the product	Often a supplier of parts, materials, etc.
Product use is	In a market niche with emphasis on its unique advantages	Expanding as more market segments are entered	Widespread
Product price	Is high and demand is insensitive to price; profit margins are high	Often failing rapidly with rising elasticity of demand	Low and demand is sensitive to price; profit margins are low
Exports are	Robust based on the product's unique performance	Strong but facing competition from imitators	Declining under strong competition from abroad
Competitors are	Few with widely fluctuating market shares	Many, but a decline in numbers begins after appearance of a dominant design	Few, a classic oligopoly situation in which market shares are stable
Unit is vulnerable to others who	Can rapidly imitate and improve on its innovations	Can produce more efficiently and consistently	Can replace its product with functionally superior or far less expensive substitutes

FIGURE 2 Hypotheses concerning the dynamics of product innovation. Reprinted with permission from Abernathy and Utterback (1978), *Technology Review*, copyright 1978.

systems become increasingly difficult to change, mechanistic, and rigid. Tasks become more specialized and are subjected to more formal operating controls. Some tasks are automated, and emphasis is placed on a systematic flow of work. Levels of automation will vary widely with "islands of automation" linked by manual operations (Bright, 1958). As a result, production processes in this state will have a segmented quality. Steps to expand capacity will most frequently include breaking bottlenecks. A larger initial investment will be required to enter the line of business during the transitional state than in the fluid state.

Having become a more significant producer and purchaser, the unit will develop suppliers that depend on its business, which will enable it to influence the consistency of its inputs. Labor tasks will gradually become more structured, with emphasis on particular skills. Maintenance, scheduling, and control will increasingly be handled by specialized labor rather than directly during production.

The production process reaches the specific state when it becomes highly developed and integrated around specific product designs, and as investment becomes correspondingly large. In this state, selective improvement of process elements becomes increasingly more difficult. Production volume and scale of plants will be large. The process becomes so well integrated that changes become extremely costly, because even a minor change may require changes in related elements of the process and in the product design. Process redesign typically comes in progressive steps, but it may also be spurred either by the development of new technology or by a sudden or cumulative shift in the requirements of the market. If changes are resisted as process technology and the market continue to evolve, then the stage is set for either economic decay or a revolutionary, as opposed to evolutionary, change.

A strong influence will be exerted on suppliers to provide consistent quality and flow of inputs, as these are critical to the unit's productivity and profits in its now high-volume and low-margin operation. Tasks that cannot be automated may be segregated from the mainstream and performed in separate locations or by subcontractors. Consequently, production scheduling and control, quality control, materials requirements planning and materials handling, job design, labor relations, and capital investment decisions will vary with changes in product and process technology.

The discussion above implies that various productivity elements—process integration, materials and labor inputs, and scale—can be considered as a set of actively coupled elements. This means that each must change in a balanced way for product and process change to advance uniformly. When one element is changing more rapidly than others, or when one or more elements are static while others are changing, we speak of start-up problems or barriers to innovation (Ramstrom and Rhenman, 1969).

Unit production costs often decrease in proportion to the cumulative volume of production. This phenomenon has been observed, for example, in the production of items as diverse as light bulbs, integrated circuits, color television sets, automobiles, and aircraft. A similar phenomenon is common in studies of individuals' performance of repetitive tasks. Because reductions in unit costs were first associated with increasing labor skills, the relationship between unit costs and cumulative production is known as a learning curve or an experience curve. Indeed, a stable and skilled labor force is apparently a prerequisite for rapid cost reduction with increases in production. The contention here is that change in each of the elements noted above, including labor skills, underlies the experience phenomenon. Further, management-determined changes in product design and process configuration make possible the required changes in other elements and thus pace the reduction in cost. These hypotheses are summarized in Figure 3.

ORGANIZATIONAL STRUCTURE AND INNOVATIVE CAPACITY

Not only do changes in products and processes occur in the systematic pattern described above, but organizational requirements may also be expected to vary according to a similar pattern (see Figure 4). During periods of high target and technical uncertainty, a productive unit must be focused to make progress; for a group to be successful in an uncertain environment, individuals in the organization must act together. This type of organizational structure is called organic (Burns and Stalker, 1961). Such an organization emphasizes, among other things, frequent adjustment and redefinition of tasks, less hierarchy, and more lateral communication. An organic organization is more appropriate to uncertain environments because of its increased potential for gathering and processing information for decision making.

The relative power of individuals in the organization will be related to their assumption of entrepreneurial roles. The rewards for radical product innovation will often be ownership of a small, new enterprise. The potential wealth resulting from such innovations will be valued by the entrepreneur to a much greater degree than his or her present income. Realization of potential rewards will depend on the survival and growth of the firm, which in turn will depend largely on the ability of the entrepreneur to generate a superior product and to capture a share of an emerging market. The innovative capacity of such an organization will be high.

As transition begins, and individuals and units in the organization be-

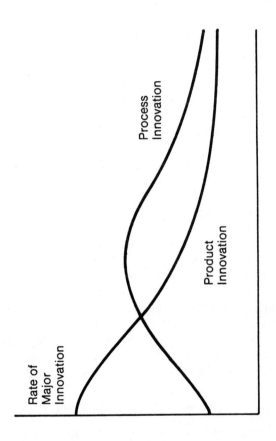

	Fluid Pattern	Transitional Pattern	Specific Pattern
Production processes	Flexible and inefficient: major changes easily accommodated	Becoming more rigid, with changes occurring in major steps	Efficient, capital-intensive, and rigid; cost of change is high
Equipment	General-purpose, requiring highly skilled labor	Some subprocesses automated, creating "islands of automation"	Special purpose, mostly automatic with labor tasks mainly monitoring and control
Process interdependence	Is slight with subprocesses being relatively independent of one another	Is rapidly increasing	Is extreme, making it difficult to incorporate changes without disrupting the rest of the process
Cost of process change	Is low	Is moderate	Is high
Materials	Inputs are limited to generally available materials	Specialized materials may be demanded from some suppliers	Specialized materials will be demanded; if not available, vertical integration will be extensive
Labor	Is highly skilled and paid and can perform a variety of tasks	Semiskilled, performs well-defined tasks at low wages	Is moderately skilled performing largely maintenance and control functions
Degree of vertical integration	Is low; the unit will purchase most of its raw materials and many parts and components	Is growing as the unit begins to produce many of its own critical parts, components, and materials	Is extensive, and usually only those units having a high degree of vertical integration will survive
Plant	Small-scale, located near user or source of technology	General-purpose with specialized sections	Large-scale, highly specific to particular products

FIGURE 3 Hypotheses concerning the dynamics of process innovation. Reprinted with permission from Abernathy and Utterback (1978), *Technology Review*, copyright 1978.

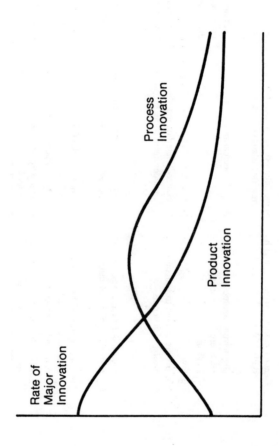

	Fluid Pattern	Transitional Pattern	Specific Pattern
Organization control is	Informal and entrepreneurial	Through liaison relationships, project and task groups	Through emphasis on structure, goals, and rules
Organizational structure is	Organic with frequent adjustment and redefinition of tasks	Hierarchical and lateral relationships are increasingly defined	Mechanistic with well-defined tasks and relationships
Requirements of management	Are entrepreneurial skills	Are increasingly the managerial skills needed to cope with growing complexity	Are those skills needed to maintain stability and moderate growth
Innovators are	Rewarded for radical product innovation, often via ownership and rapid growth of the unit	Rewarded for expansion of operations and contributions to rapid productivity gains	Discouraged from pursuing ideas that threaten the stability of the unit
Innovative capacity of the organization	Is high	Is moderate	Is low due to the disruptive nature of major innovations

FIGURE 4 Hypotheses concerning the dynamics of organizational form. Reprinted with permission from Abernathy and Utterback (1978), *Technology Review*, copyright 1978.

come more sequentially interdependent, coordination and control will occur to a greater extent through planning, liaison relationships, and project and task groups. When the task environment is better defined, an appropriate means of coordination and control will be to define various lateral relationships based on the dependence of one part of the organization on another (Lawrence and Lorsch, 1967). Thus, during transition, organizations are often structured according to products, or regions, each division replicating in some respects the earlier entrepreneurial form.

The relative power of individuals will begin to shift from those with entrepreneurial ability to those with management ability, for a different set of skills will be required for the growth and structuring of the organization. Often the original entrepreneur or entrepreneurial group will spin off to start another smaller enterprise.

As a dominant design emerges and production operations expand rapidly in response to increased demand, the focus of rewards will shift to those who are able to expand production operations, marketing functions, and so forth. Ownership of the unit by this time may be well established, and rewards may be provided in more traditional terms of bonuses, stock options, and other managerial prerequisites.

These changes will cause moderation of the innovative capacity of an organization. As a product becomes more standardized and is produced in a more systematized process, interdependence among organizational subunits gradually increases, making it more difficult and costly to incorporate radical innovations.

Once a production process becomes highly developed with respect to a specified and standardized product, organizational control will be provided through emphasis on structure, goals, and rules. When the environment is better known and operations become routinized, it is necessary to provide more rigid coordination that establishes consistent routines and rules to minimize inefficiency and costs in operations. This type of structure is known as mechanistic.

The power and influence of individuals who show administrative ability will increase in a mechanistic organization. When the technical and market environment becomes stable—and when growth of a productive unit relies more on stretching existing products and processes—the ability to hold a steady and consistent course will be highly valued.

Rewards in a stable situation will be centered on financial results and on predictable, incremental performance in product and process change that build on past investments. Ideas that threaten to disrupt the stability of the existing process will be discouraged, and ideas that extend the life of existing products and technology will be encouraged and rewarded, probably in a highly structured manner.

The innovative capacity of such a productive unit viewed in isolation will be low. When production processes are highly integrated in a system, and a high degree of interdependence exists among subprocesses, the disruption and cost associated with major changes will be a primary concern. Moreover, influential individuals' perceptions of the gains from improvements that provide immediate and certain rewards amplified by a high production volume will be clear, whereas the consequences of changes that are costly, uncertain, or delayed may be greeted with great skepticism. These hypotheses are summarized in Figure 4.

These issues raise difficult problems for organizations. In an organization with a diversity of products in different markets and at different phases in the dynamic product cycle, there is a serious problem of fitting together the organizational styles required for each of the stages. A subdivision that may be the logical functional locus for the introduction of a new product because of similarity of market may have a hierarchical, bureaucratic organization more appropriate to a mature old product and therefore be unable to accommodate itself to the innovation.

TRANSITION FROM RADICAL TO EVOLUTIONARY INNOVATION

Although we have discussed the three different patterns of innovation as distinct modes of change, they are not completely rigid and independent. That is, each pattern has definable characteristics, but the lines between them tend to blur in real life. Several examples illustrate how movement from one pattern to another proceeds.

John Tilton's study (1971) of developments in the semiconductor industry from 1950 through 1968 indicates that the rate of major innovation decreased and that the type of innovation shifted. Eight of 13 major product innovations occurred in the first 7 years of that period, during which time the industry was making less than 5 percent of its total 18-year sales.

Two types of enterprise can be identified in this early period—established units that came to semiconductors from vested positions in vacuum tube markets, and new entrants such as Fairchild Semiconductors, IBM Corporation, and Texas Instruments Inc. The established units responded to competition from the newcomers by emphasizing process innovations, whereas the newcomers sought entry and strength through product innovation. The three successful new entrants just listed were responsible for half of the major product innovations and only one of the nine process innovations Tilton identified in that 18-year period; however, the three principal established units (divisions of General Electric, Philco, and RCA) made only one-quarter of the product innovations in the same period. Here, process innovation did not prove to be an effective competitive stance; by

1966, the three established units together held only 18 percent of the market, whereas the three new units held 42 percent. Since 1968, however, the basis of competition in the semiconductor industry has changed; as costs and productivity have become more important, the rate of major product innovation has decreased, and effective process innovation has become an important factor in competitive success.

Like the transistor in the electronics industry, the DC-3 stands out as a major change in the aircraft and airlines industries. Almarin Phillips (1971) has shown that the DC-3 was a culmination of previous innovations. It was not the largest, fastest, or longest-range aircraft; it was the most economical, large, fast plane able to fly long distances. It was also essentially the first commercial product of an entering firm (the DC-1 and DC-2 were produced by Douglas in only small numbers).

The DC-3 changed the character of innovation in the aircraft industry. No major innovations were introduced into commercial aircraft design from 1936 until jet-powered aircraft appeared in the 1950s. Instead, there were many incremental refinements to the DC-3 concept, which lowered airline operating costs per passenger-mile an additional 50 percent.

The history of the electric light bulb also shows a series of evolutionary improvements that started with a few major innovations and ended in a highly standardized commodity-like product (Bright, 1949). By 1909, the first tungsten filament and vacuum bulb innovations were in place; from then until 1955 there came a series of incremental changes that dropped the price of a 60-watt bulb from $1.60 to $0.20 (even with no inflation adjustment), increased the lumens output by 175 percent, and reduced the direct labor content from 3 to 0.18 minutes per bulb. The production process evolved from a flexible job-shop configuration, with more than 11 separate operations and a heavy reliance on the skills of manual labor, to a single machine attended by a few workers.

Product and process evolved in a similar way in the automobile industry (Abernathy, 1978). During the 4-year period from 1905 to 1909, the Ford Motor Company developed, produced, and sold five different engines, ranging from two to six cylinders. Each engine tested a new concept. They were made in a factory that was flexibly organized, much as a job shop, relying on trade craftsmen working with general-purpose machine tools that were not the best then available. Out of this experience came a dominant design—the Model T—in 1909, and within 15 years, 2 million engines of this design were produced each year (about 15 million in all) in a facility then recognized as the most efficient and highly integrated in the world. During that 15-year period there were incremental, but not fundamental, innovations in the Ford product.

The shift from radical to evolutionary product innovation is a common thread in these examples. It is related to the development of a dominant product design, and it is accompanied by heightened price competition and increased emphasis on process innovation. Small-scale units that are flexible and highly reliant on manual labor and craft skills using general-purpose equipment develop into units that rely on automated, equipment-intensive, high-volume processes. Thus, changes in innovative pattern, production process, and scale and kind of production capability all occur together in a consistent, predictable way.

DYNAMICS OF A SET OF COMPETING PRODUCTIVE UNITS

Creative synthesis of a new product innovation by one or a few firms may result in a temporary monopoly, high unit profit margins and prices, and sales of the innovation in those few market niches where it possesses the greatest performance advantage over competing products. As volume of production and demand grows, and as a wider variety of applications is opened for the innovation, many new firms enter the market with similar products.

The appearance of a dominant design shifts the competitive emphasis to favor those firms with a greater skill in process innovation and process integration, and with more highly developed internal technical and engineering skills. Many firms will be unable to compete effectively and will fail. Others may possess special resources and thus merge successfully with the ultimately dominant firms, whereas weaker firms may merge and still fail.

Eventually, the market reaches a point of stability, corresponding to the specific state, in which there are only a few firms—four or five is a typical number from the evidence reviewed to date—having standardized or slightly differentiated products and stable sales and market shares. A few small firms may remain in the industry, serving specialized market segments, but, as opposed to the small firms entering special segments early in the industry, they have little growth potential. Thus, it is important to distinguish between small surviving firms and small firms that are new entrants, and to keep in mind that the term "new entrants" includes existing larger firms moving from their established market or technological base into a new product area.

Mueller and Tilton (1969) were among the first to present this hypothesis in its entirety. They contend that a new industry is created by the occurrence of a major process or product innovation and develops technologically as less radical innovations are introduced. They further argue that the large corporation seldom provides its people with incentives to introduce a development of radical importance; thus, these changes tend to be developed by new entrants without an established stake in a product market segment. In their words, neither large absolute size nor market power is a necessary condition for successful competitive development of most major innovations.

Mueller and Tilton contend that once a major innovation is established, there will be a rush of firms entering the newly formed industry or adopting a new process innovation. They hold that during the early period of entry and experimentation immediately after a major innovation, the science and technology on which it depends are often only crudely understood, and that this reduces the advantage of large firms.

However:

> As the number of firms entering the industry increases and more and more R&D is undertaken on the innovation, the scientific and technological frontiers of the new technology expand rapidly. Research becomes increasingly specialized and sophisticated and the technology is broken down into its component parts with individual investigations focusing on improvements in small elements of the technology [Mueller and Tilton, 1969, p. 576].

This situation clearly works to the advantage of larger firms in the expanding industry and to the disadvantage of smaller entrants.

Staples, Baker, and Sweeney (1977) have summarized several clear parallels between the present theory and Mueller and Tilton's hypotheses:

> The Utterback and Abernathy model holds implications for organizational structure, just as Mueller and Tilton's does for the composition of an industry. A comparison of the two will show a number of similarities. Both describe a continuum. The stages roughly correspond. Both emphasize the shift of the basis of competition from performance and technological characteristics to price and cost considerations. In both, the evolution is accompanied by an expanding market, increasing importance of production process investment, and a progression from radical to incremental product and process innovation. In general, they describe a progression from a state of flux with rapid technological progress to an ordered situation with cumulative incremental changes. Although they emphasize different aspects of innovation from different perspectives, the models are consistent [Staples et al., 1977, p. 12].

Both this work and that of Mueller and Tilton contend that as an industry stabilizes—that is, as technological progress slows down and production techniques become standardized—barriers to entry increase. The most attractive market segments will already be occupied. As process integration progresses, the cost of production equipment rises dramatically. Product prices will fall concurrently, so that firms with the largest market shares will be the ones to benefit from further expansion. Product differentiation will usually be increasingly centered around the technical strengths and R&D organization of the existing firms. Strong patent positions established by earlier entering firms become difficult for later entrants to circumvent.

Finally, an existing distribution network may also be a powerful barrier to entry, particularly to foreign firms.

Another hallmark of stability is the emergence of a set of captive suppliers of equipment and components. Although such suppliers can be an initial source of innovation and growth, they ultimately may become a conservative force, further stabilizing the competition and change within the product market segment, and creating yet another barrier to entry.

A final characteristic of the evolution toward stability is a concerted drive among the surviving firms toward vertical integration from materials production to sale. This integration may take various forms. Firms producing the product can reach backward to furnish more of their own components, subassemblies, and raw materials, or the firms producing components can reach forward to do more of the assembly and production of final goods for the market. Such dramatic changes will clearly have ripple effects on firms that buy from or sell to the evolving set of productive units. It is just at the point of stability in which firms get locked into narrow positions that they also ultimately increase their vulnerability. An existing distribution network can suddenly be threatened by a new technology that requires sharply reduced servicing or maintenance, or by the entrance of a large product line. An existing patent may expire. Although Mueller and Tilton contend that industries become stable when patent positions expire, the present argument is that this is more likely to be a period of invasion of the industry by a new wave of product and process change—or, in a few cases, the revitalization of the dominant technology itself.

The development of a set of productive units is expected to begin with a wave of entry gradually reaching a peak about when the dominant design of the major product emerges, and then rapidly tapering off. This sequence is followed by a corresponding wave of firms exiting from the industry. The sum of the two waves—entries and exits—will yield the total number of participants in the product market segment at any point. Therefore, the number of participants in an industry can be represented by a curve that starts with a gentle rise representing the first few fluid productive units entering the business followed by a much sharper rise that represents a wave of imitating firms. The point at which a dominant design is introduced in the industry is followed by a sharp decline in the total number of participants until the curve of total participants reaches the stable condition with a few firms sharing the market.

We will now turn to a discussion of the auto industry, which exhibits such a wave of change.[1] We will then turn to the question of successive waves of entry, and again present specific examples that illustrate the applicability of the model.

THE AUTOMOBILE

More than 100 firms entered and participated in the American automobile industry for a period of 5 years or longer. Figure 5 shows the wave of entry that began in 1894 and continued through 1950, followed by a wave of exits beginning in 1923 and peaking only a few years later, although it has continued until the present day.

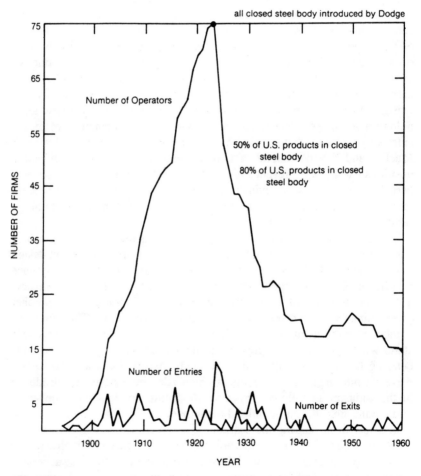

FIGURE 5   Entry and exit of firms in the U.S. automobile industry: 1894-1962. Data from Fabris (1966).

As hypothesized, entry began slowly but then accelerated rapidly after 1900, reaching a peak of 75 participants in 1923. In the next 2 years, 23 firms, nearly a third of the industry, left or merged, and by 1930, 35 firms had exited. During the ensuing depression, 20 more firms left.[2] There was a brief flurry of entries and then exits immediately after World War II, but as Figure 5 shows, the number of firms in the industry was relatively stable from 1940 through 1960.

The number and scope of major product innovations are reflected in this pattern of entries and exits. In 1923, the year with the largest number of firms, Dodge introduced the all-steel, closed-body automobile. The large number of exits over the next few years corresponds to the fact that by 1925, 50 percent of United States production was closed steel-body cars, and by 1926, 80 percent of all automobiles were of this type. The post-World War II stability in market shares and number of firms reflects the fact that approximately three-quarters of the major product innovations occurred before the start of the war.[3]

Innovations in product accessories and styling concepts were tested in the low-volume, high-profit luxury automobile. Conversely, incremental innovations were more commonly introduced in lower-price, high-volume product lines. General Motors led in both types of innovations, particularly for major product changes. In certain years, engines show a higher annual magnitude of changes; these changes, however, occur with less frequency than those in chassis characteristics; body productive units are more flexible and continuously changing than engine plants, which tend to change occasionally in an integrated and systematic way.[4]

COMPARATIVE ANALYSIS

As a productive unit develops, its reliance increases on outside sources for production process equipment and components. Firms in the auto industry, for example, developed an early and increasing reliance on suppliers for many types of equipment and innovations (Abernathy, 1978, pp. 60-61).

Development of relationships with suppliers, and of a captive set of suppliers, is a hallmark of all the evolving product market segments cited. For example, during the 1890s George Eastman was helped greatly by the availability of high-quality papers and chemicals, some of which had been developed for the earlier dry-plate photographic market; he was also assisted by the rapid increase in the number of firms manufacturing cameras. Several such firms were subcontracted by Kodak to manufacture camera backs and shutters to Eastman's design. Similarly, during the 1970s the large number of companies assembling electronic calculators were greatly aided by the availability of high-quality components with declining prices from semiconductor manufacturers. However, in both these examples, these strengths eventually became weaknesses. Established competitors tried to bankrupt Kodak by capturing the two or three sources of high-quality photographic paper, thus drying up his supply. Then, unexplained quality variations in the celluloid that he purchased combined with other circumstances to make several months' production of film useless. Finally, financial weakness and instability among the various firms manufacturing cameras threatened to make it difficult and expensive for Kodak to provide cameras to customers. These events pushed Eastman first in the direction of producing its own photographic paper, then its own chemicals, and finally its own cameras, camera backs, and shutters (Bright, 1949; Jenkins, 1975).

Advantages also turned to disadvantages in the calculator industry when rapid reductions in the price of semiconductors caused enormous inventory losses for firms that were purely assemblers. As the production capacity of semiconductor manufacturers increased and production costs dropped further, virtually all the value-added in the calculator occurred in the manufacture of its components; these firms simply integrated forward to provide the entire calculator for the user (Majumdar, 1977). Thus, while suppliers may play a highly creative role as a set of productive units develops, there will also be a drive among producing firms to capture those elements of supply that create the greatest uncertainties for them.

However, it should be pointed out that the most innovative producers always seem to provide some of their own production equipment. Abernathy (1978) and Fabris (1966) show that General Motors and, especially, Ford have made continuing process innovations. In the semiconductor industry, Texas Instruments, in particular, has stressed production process innovation and integration, and Tilton's data (1971) show a pronounced shift toward process innovation by new firms as the industry developed and as their market shares expanded.

In summary, those firms that survive the introduction of a dominant design appear to be those that integrate vertically or establish the closest supplier relationships. Sometimes, it is the supplier who integrates forward, rather than an early manufacturer who integrates backward, that dominates the market. As Abernathy observed:

> The degree of vertical integration is not static as long as major product changes are taking place. It is rather the equilibrium condition of a continuous effort to extend integration backward in the face of the constant erosion caused by product change. As the technology of product design advances so that novel changes [are] made less necessary, vertical integration can be maintained without such continuous effort [Abernathy, 1978, pp. 110-111].

Examining market structure during waves of change indicates that firms that are highly integrated are the most vulnerable to functional technological competition, for they have developed stable production processes and sources of supply and thus have a major commitment to the existing technology (Utterback and Kim, 1986). They may view the new technology or product as either highly specialized with a narrower and small market or as an inferior good, also with narrow market appeal. For example, all the major vacuum tube firms adopted the transistor for traditional applications of tubes. Gene Strull of Westinghouse Electric Corporation has been quoted by Braun and MacDonald (1978) to the effect that probably every major older company began the use of transistors in the divisions that had been making tubes for the same purposes. Strull claims that this practice handicapped the introduction of semiconductors because it made them look like a replacement for the tube; it was a few years before people started to look and see what the transistor could do in its own right.

All major mechanical calculator firms were early entrants in the electronic calculator business. However, they emphasized the complexities of the electronic calculators, and produced them only for the most difficult and limited scientific applications, not in broader and simpler lines for use in business (Majumdar, 1977).

The major mechanical typewriter firms were early entrants in the manufacture of electric typewriters but did not continue with their innovations after World War II. The government played a role here in that it directed the typewriter companies to manufacture various types of arms for the war effort and specifically enjoined them from making typewriters. Since IBM Corporation was not in a critical labor supply area, it was allowed to continue manufacturing electric typewriters, nearly all of which were placed in government and military offices. This not only allowed IBM to expand its technical capability and market share, but it also introduced a wide variety of people to use of the new electric typewriter. When IBM's competitors reentered the new business after the war, they all did so with their traditional mechanical designs (Engler, 1965).

Finally, companies producing woven carpets of wool were placed at a double disadvantage by the innovation of tufted carpeting using synthetic fibers. Firms producing woven woolen carpets had strong ties with wool suppliers and controlled, through purchases, nearly the entire wool market. Synthetic materials not only enabled the new tufting technology to be highly productive, but they allowed the carpet market to expand dramatically with falling, rather than rising, marginal costs, an experience that was foreign to the manufacturers of woven carpets (Reynolds, 1967).

The previous examples have shown how firms enter and leave an industry in parallel with product innovation in that industry. In the fluid state, while product requirements are still ambiguous, there will tend to be a rapid entry of firms and few failures. As the industry enters the transitional state, and product requirements become more defined, fewer firms enter and a larger number either may merge or fail. Finally, as the industry enters the specific state, there are only a few large firms, each controlling a consistent share of the market, and possibly a few small firms serving highly specialized segments.

INNOVATION, ORGANIZATIONAL STRUCTURE, AND INTERNATIONAL TRADE

The literature on technology and international trade has shown that shifts in innovation and industry structure are tied to shifts in the location of production and flows of trade. Louis Wells (1972) finds that trade varies with the product life cycle as follows: Innovation occurs and production begins in the country with the largest and most demanding market for a product—typically the United States. Exports quickly begin to serve mid-scale markets—Europe and Japan. Production then begins in the early export markets with the focus of exports starting to shift to less well developed markets, such as South America. Europe and Japan begin exporting to developing countries in competition with the United States while manufacturing begins in those countries also. Ultimately, producers in developing nations begin exporting back to Europe, Japan, and the United States. An essential point of this argument is that once a product becomes a commodity and the technology stabilizes—that is, when it enters the specific state—maintaining control of production becomes increasingly difficult. This is especially so if other countries have great advantages in factor costs, including materials and energy as well as labor. Conversely, if technology is rapidly changing, innovation and manufacture are much more likely to occur close to users. Freeman (1968) has linked this phenomenon to the export of process equipment. Hekman (1980) has shown that the rapid advance of textile technology led manufacturers to cluster around Boston in the 1830s. He further shows that as production technology stabilized, the industry became widely dispersed in part through the export of now-standardized textile equipment from Boston.

Linsu Kim (1980) has pursued a similar hypothesis in reverse in the contemporary and international setting of the Korean electronics industry. He found that the industry became established in Korea through transfers of standardized technology to firms having both the strategy and the organization capable of absorbing it. Later, these firms began to produce variations in product and process. The learning and adjustment engendered by the firms' incremental innovations help to create an organization that may become "innovation viable," that is, an organization able to succeed in making product changes in larger steps and to compete on a more equal and independent footing in export markets.

Nearly all these examples point to the hypothesis that entering early is the most viable strategy for a firm. If we based our assessment of technological and market dynamics strictly on U.S. business history, it would be hard to disprove this hypothesis. For example, carbon-filament incandescent lamps replaced gas lighting; they themselves were replaced by metal-filament incandescents and later by fluorescent lighting. The Edison Company and the Swan Lamp Company were the innovators in carbon-filament lamps, but only an insurmountable patent position in other aspects of lamp manufacture allowed Edison to overcome new firms that adopted metal filaments earlier than it did. Sylvania in the United States was the first to innovate with fluorescent lighting, and it increased its market share from 5 to 20 percent at General Electric's expense. Harvested, naturally formed ice for refrigeration was replaced by machine-made ice and later by mechanical refrigeration; it was not the ice-harvesting companies that innovated in mechanical means of ice production, nor was it the companies producing ice and ice boxes who innovated in the area of electromechanical refrigeration. Finally, in the 20 years from 1889 to 1909, Eastman Kodak's share of the U.S. photographic market went from 16 percent to 43 percent at the expense of established makers of dry photographic plates, because of its innovation of celluloid roll film.

Whereas some investigators of technology and corporate strategy in the United States have emphasized the value of early entry with an innovation, Harvey Brooks writes:

> The typical pattern of Japanese success has been rapid penetration of a narrow, but carefully selected segment of a broad, expanding world market in which superiority in production efficiency, economies of scale, and exploitation of learning curve effects were particularly important. By expanding more aggressively than its U.S. competitors and anticipating learning curve improvements and economies of scale further into the future in its pricing strategies, Japan has been able to capture an important share of the market for selected products just behind the current technological frontier. They have then broadened out from this point in the middle technology spectrum and moved gradually toward more sophisticated and higher value-added products in the same or a closely allied market segment. Willingness to plunge in and adopt a new technology on the basis of its ultimate promise before it was proven to be cost-effective has been combined with careful and thorough scanning of related world technological developments for their possible competitive threat or promise [Brooks, 1985, p. 330].

The success of Japanese firms in U.S. markets for automobiles and steel raises a variety of questions about business strategies in technologically dynamic product markets. Clearly, in the past each wave of radical product change has brought with it the entry of new firms—either small, technology-based enterprises or larger firms carrying their technical skills into the new product and market areas—and these firms may dominate the restructured industry. The Japanese example, however, shows that productive units can pursue widely different strategies as long as the strategy is matched to the state of evolution of the technology. Clearly, the dynamics of technological change in relation to corporate strategy and international competition are fruitful areas for further work. This is especially so in the light of changing organizational forms and the increasing integration of production across national boundaries, issues discussed by Doz and Teece in subsequent chapters in this volume.

SUMMARY

In summary, to understand how the development and diffusion of technology affects national productivity and competitiveness, it is essential that we understand the linkages of product technologies with manufacturing process, corporate organization and strategy, and the structure and dynamics of an industry. Lacking balance and integration among all essential factors means that by investing heavily in one area, a firm could allow its competitors to exploit the new product or process technology first.

Focusing on manufacturing (or product development, or finance) alone is wholly insufficient. Product design for manufacture, change in organization, and appropriate strategy are also prerequisites for competitive strength. By the same token, potential for product innovation and competitiveness depends increasingly on ability to innovate in manufacturing processes. Finally, there exists a hierarchy of productive units—a product for one is part of the process for another and therefore affects productivity directly. Productivity at the final use stage is strongly affected by the vitality of productive units at earlier stages. Lack of responsive suppliers of equipment and components will seriously constrain advances in ultimate products and systems. Moreover, it is not clear to what degree a nation can import process equipment and assume that its long-run competitive and innovative strengths will not be eroded.

With regard to industry structure, appearance of a dominant design shifts emphasis to manufacturing for survival. Those who fail to shift will usually not survive. The dominant design should address world markets and standards to be most competitive (see Lehnerd in this volume). Similarly, it is a mistake in competition to automate too soon or too extensively. Doing so may reduce flexibility in the face of continuing product change and may leave a firm with heavily capitalized plants that are obsolete the day they come on stream. Tailored manufacturing approaches that allow needed flexibility in product characteristics are often hallmarks of the most successful and competitive firms.

As a product design stabilizes, diffusion of technology is inevitable through movement of skills and people as well as advanced equipment. Architectural rebuilding of industry is constantly required; a vital area for research is the discovery of means through which large and established organizations can constantly and creatively renew their businesses. In an organization with a diversity of products in different markets and at different phases in the dynamic product cycle shown in Figure 1, there is a serious problem of fitting together the organizational styles required for each of the different stages. A subdivision that may be the logical functional locus for the introduction of a new product because of similarity of market may have a hierarchical, bureaucratic organization more appropriate to a mature old product and therefore be unable to accommodate itself to the innovation. This may have been one of the main reasons why vacuum tube divisions that initially seemed to be at the forefront of transistor and semiconductor technology (where they benefited from government support) were unable to become the leaders in the market for this technology when it moved from the fluid stage to the transitional stage. Purely entrepreneurial strategies may no longer be sufficient for successful entry. Rather, creative coalitions blending the strengths of both new and established firms may be required for success in a more international competitive arena.

ACKNOWLEDGMENTS

I am especially indebted to the late William J. Abernathy. Our collaboration led to many of the ideas and findings expressed here. Many others were originated by him and are explored in the context of the auto industry in his book *The Productivity Dilemma* (Baltimore: Johns Hopkins University Press, 1978). This report is based on work supported by the National Science Foundation, Division of Policy Research and Analysis under Grant No. PRA76-82054 to the Center for Policy Alternatives at the Massachusetts Institute of Technology.

I also owe a special debt to both Harvey Brooks and Bruce Guile. The original manuscript for this was written in 1982 as part of the above-mentioned project. Harvey Brooks provided an extensive and challenging commentary on the manuscript. Many of his questions are addressed in part here, much improving the resulting document, but many remain to be addressed. Bruce Guile helped far beyond any reasonable expectation not only in thoroughly editing the manuscript but in providing essential suggestions, advice, and encouragement.

NOTES

1. Many other examples also could be cited to support these hypotheses. For example, Arthur Bright's work (1949) on invention and innovation in the electric lamp industry cites detailed statistics on firms' entrances and exits, and he gives elaborate histories of the major firms—Westinghouse, the Thompson-Huston Company, and the Edison Company, the latter two of which later merged to become General Electric. Phillips (1971) and Miller and Sawers (1970) provide similar data on air frame and aircraft engine manufacturers; these data have been summarized in another paper by Linsu Kim (1980). Anderson (1953) gives general figures on the number of participants in different phases of the American ice and refrigeration industry, and Jenkins (1975), while concentrating on the Eastman Kodak Company, also discusses the formation, merges, and demise of many other competing firms.
2. The material in this section is based on a dissertation by Richard Fabris entitled "Product Innovation in the Automobile Industry," written in 1966. Supplementary information on the origin and diffusion of different major innovations has been obtained from William Abernathy's book, *The Productivity Dilemma*, and on market shares and entry from Burton H. Klein's book, *Dynamic Economics*.
3. These figures are somewhat understated because Fabris does not count a firm that merged but continued in a larger conglomerate as leaving the industry—for example, Cadillac and the Oakland Company (now Pontiac) are counted as surviving independent firms.
4. Fabris studied 32 major product innovations and found that 70 percent occurred before 1935. Abernathy (1978) includes three additional major innovations as occurring during this period—the aluminum alloyed piston, the automatic choke, and disc brakes. Two more of Abernathy's major product innovations—energy absorbing steering assemblies and 12-volt electrical systems—follow the 1962 termination of Fabris' analysis, so there is about a two-thirds overlap between the two studies.

REFERENCES

Abernathy, W. J. 1976. Production process structure and technological change. Decision Science 7 (October):607-619.

Abernathy, W. J. 1978. The Productivity Dilemma. Baltimore, Md.: Johns Hopkins University Press.

Abernathy, W. J., and P. L. Townsend. 1975. Technology, productivity and process change. Technological Forecasting and Social Change 7(4):379-396.

Abernathy, W. J., and J. M. Utterback. 1978. Patterns of innovation in technology. Technology Review 80:7(June-July):40-47.

Abernathy, W. J., and K. Wayne. 1974. Limits of the learning curve. Harvard Business Review 52(5):109-119.

Anderson, O. E., Jr. 1953. Refrigeration in America: A History of a New Technology and Its Impact. Princeton, N.J.: Princeton University Press.

Braun, E., and S. MacDonald. 1978. Revolution in Miniature: The History and Impact of Semiconductor Electronics. Cambridge, England: Cambridge University Press.

Bright, A. A., Jr. 1949. Electric Lamp Industry: Technological Change and Economic Development from 1800 to 1947. New York: MacMillan.

Bright, J. R. 1958. Chapter 1 in Automation and Management. Division of Research, Graduate School of Business Administration. Boston: Harvard University.

Brooks, H. 1985. Technology as a factor in competitiveness. Pp. 328-356 in U.S. Competitiveness in the World Economy, B. R. Scott and G. C. Lodge, eds. Boston, Mass.: Harvard Business School Press.

Burns, T., and G. M. Stalker. 1961. The Management of Innovation. London: Tavistock.

Engler, G. N. 1965. The Typewriter Industry: The Impact of a Significant Technological Innovation. Ph.D. dissertation. University of California, Los Angeles.

Fabris, R. 1966. Product Innovation in the Automobile Industry. Ph.D. dissertation. University of Michigan.

Freeman, C. 1968. Chemical process plant: Innovation and the world market. National Institute Economic Review 45:29-51.

Frischmuth, J. S., and T. J. Allen. 1969. A model for the description of technical problem solving. IEEE Transactions on Engineering Management EM-12 (May):79-86.

Hekman, J. S. 1980. The product cycle and New England textiles. Quarterly Journal of Economics 94(4):697-717.

Hill, C. T., and J. M. Utterback, eds. 1979. Chapter 2 in Technological Innovation for a Dynamic Economy. Pergamon Press.

Jenkins, R. V. 1975. Images and Enterprise: Technology and the American Photographic Industry, 1839 to 1925. Baltimore, Md.: Johns Hopkins University Press.

Kim, L. 1980. Stages of development of Industrial Technology in a developing country: A model. Research Policy 9 (July):154-177.

Klein, B. H. 1977. Dynamic Economics. Cambridge, Mass.: Harvard University Press.

Lawrence, P. R., and J. W. Lorsch. 1967. Organization and Environment. Division of Research, Harvard Business School. Boston: Harvard Business School.

Majumdar, B. A. 1977. Innovations, Product Developments, and Technology Transfers: An Empirical Study of Dynamic Competitive Advantage, The Case of Electronic Calculators. Ph.D. dissertation. Case Western Reserve University.

Miller, R. E., and D. Sawers. 1970. The Technical Development of Modern Aviation. New York: Praeger Publishers.

Mueller, D. C., and J. E. Tilton. 1969. R&D cost as a barrier to entry. Canadian Journal of Economics 2 (November):576.

Normann, R. 1971. Organizational innovativeness: Product variation and reorientation. Administrative Science Quarterly 16 (June):203-215.

Phillips, A. 1971. Technology and Market Structure: A Study of the Aircraft Industry. Lexington, Mass.: Heath Lexington Books.

Ramstrom, D., and E. Rhenman. 1969. A method of describing the development of an engineering project. IEEE Transactions on Engineering Management EM-16 (May):58-64.

Reynolds, W. A. 1967. Innovation in the U.S. Carpet Industry, 1947-1963. Ph.D. dissertation. Columbia University.

Rosenbloom, R. S. 1974. Technological innovation in firms and industries: An assessment of the state of the art. Harvard Business School Working Paper. HBS 74-8. Boston: Harvard Business School.

Staples, E. P., N. R. Baker, and D. J. Sweeny. 1977. Market Structure and Technological Innovation: A Step Towards a Unifying Theory. Final Technical Report. NSF Grant RDA 75-17332, November.

Tilton, J. E. 1971. International Diffusion of Technology: The Case of Semiconductors. Washington, D.C.: The Brookings Institution.

Utterback, J. M. 1975. Innovation in industry and the diffusion of technology. Science 183:620-626.

Utterback, J. M. 1978. Management of technology. Pp. 137-160 Studies in Operation Management, Arnoldo Hax, ed. Amsterdam: North Holland.

Utterback, J. M., and W. J. Abernathy. 1975. A dynamic model of process and product innovation. Omega 3(6):639-656.

Utterback, J. M., and L. Kim. 1986. Invasion of a stable business by radical innovation. Pp. 113-151 in The Management of Productivity and Technology in Manufacturing. New York: Plenum Press.

Vernon, R. 1966. International investment and international trade in the product cycle. Quarterly Journal of Economics 80(2):190-207.

von Hippel, E. 1977. The dominant role of the user in semi-conductor and electronic subassembly process innovation. IEEE Transactions on Engineering Management EM-24 (May):60-71.

Wells, L. T. 1972. The Product Life Cycle and International Trade, Division of Research. Boston, Mass.: Harvard Business School.

White, G. R. 1978. Management criteria for effective innovation. Technology Review 80(4):14-22.

Revitalizing the Manufacture and Design of Mature Global Products

ALVIN P. LEHNERD

Manufacturing enterprises are evolutionary entities. Over time, their product portfolios expand through evolutionary and chronological developments. Products are usually designed and developed one at a time. As a result, it is the exception when the designs of a manufacturer's products embrace much compatibility, standardization, or modularization. The norm is that product portfolios are rarely designed simultaneously; designs take place in a sequential manner. Additionally, many current products of U.S. manufacturers were designed and tooled years ago, yet prevailing labor rates, manufacturing processes, energy costs, availability of materials, and interest rates are often significantly changed from the time of the original product design and tooling activities. It is rare that a U.S. manufacturer invests the time and resources necessary to rationalize production of an entire product line to fit the changing economic environment and to take advantage of opportunities provided by technological advance.

Manufacturers usually design for function, then redesign for manufacturing; thus, two design iterations usually take place. If an enterprise wishes to maintain or gain market share in global markets, the firms' managers and technical personnel must learn to combine manufacturing with innovation in product design. Few enlightened companies take time for a third design iteration to automate and mechanize production for global leadership in cost and value.

In many, if not all, instances, design for manufacturing is also constrained by the existing resources of plant and equipment. In other words, manufacturing engineers guide the design decisions to match the profiles

49

and capabilities of their existing factories and their respective in-house capabilities. In-place facilities are frequently barriers to product innovations. Fixed capital investments in existing capabilities are also barriers to more advanced lower-cost processes. Organizations commonly ignore what the production cost could be if their products were not shackled to outdated manufacturing processes and could also use state-of-the-art materials requiring new processes and procedures.

An additional issue is that few U.S. domestic manufacturers look at their product offerings as global opportunities. This domestic myopia—the belief that the marketplace ends at the U.S. borders—is a problem for U.S. industry, and the problem will only get worse as the world becomes more economically integrated.

Finally, corporate planning horizons are too short, and manufacturers seldom ask themselves what they are doing to ensure their longevity in the business. Too many managements or boards of directors do not act until external influences cause significant disruptions and spur the organization into action.

This chapter presents a case history of a 1970s program at Black & Decker Corporation to redesign a product line for production automation and leadership in cost and value. The program was an effort to redesign standard products to take advantage of opportunities for using new materials and new manufacturing and design techniques.

BLACK & DECKER

When managers at Black & Decker Corporation observed growing global competition in the 1960s and 1970s, they decided that a window of opportunity existed to improve their product lines and manufacturing capability. Moreover, they decided that if they did not take time to do it right the first time, they would never have the time or resources to do it over. They recognized that if they were to be a domestic manufacturer with aspirations to do business internationally 20 years hence, they would have to change the business in a way that would ensure that long-range performance. This involved making certain irrevocable decisions.

The impetus for change came from three sources. First, it was evident that foreign competition would increase in Black & Decker's product markets and that this would lead to foreign participation in new, related product markets.

Second, in the 1970s, inflation in costs of labor, material, services, and capital goods was a serious consideration. Table 1 shows the effect of inflation in the labor component of product costs. It assumes an 8 percent compounded inflation rate over five periods from year 1 to year 6. To

TABLE 1   Impact of Wage Inflation on Labor Costs (8 percent compounded inflation)

Year	Hourly Wage ($)	Labor Minute Value of $3.00
1st	3.00	60.0
2nd	3.24	55.5
3rd	3.50	51.5
4th	3.78	47.6
5th	4.08	44.1
6th	4.41	40.8

maintain constant labor-cost content in the product, one-third of the labor has to be removed from the product between period 1 and period 5. In Black & Decker's assessment, offsetting inflation in labor costs depended on making better use of labor in adding value through design standardization, mechanization, automation, better use of material and floor space, and intelligent capital planning.

The third factor in Black & Decker's decisions was an anticipated continued public attention to consumer protection and environmental concerns. In the power tool industry, this attention took the form of requirements for double insulation of tools. The term "double insulation" refers to the additional insulation barrier placed in an electrical device to protect the user from electrical shock if the main insulation system ever fails. In the late 1960s there was a strong possibility that double insulation of domestic power tools would be legally required. Black & Decker decided that the threat of required double insulation provided an opportunity to study the entire product line.

The program begun at Black & Decker in the early 1970s was called Double Insulation. All consumer power tools were to be redesigned at the same time and would initially be manufactured in various locations in North America with standardized parts and components.

Double Insulation was Black & Decker's vehicle to redesign the line and develop a "family" look, simplify the product offering, reduce manufacturing costs, automate manufacturing, standardize components, incorporate new materials, improve product performance, incorporate new product features, and provide for worldwide product specifications. The program was designed to incorporate double insulation on 122 basic tools with hundreds of variations. Of 18 tool groups manufactured by Black & Decker, 8 contributed 73 percent of sales, 71 percent of earnings, and 91 percent of units sold. These groups were tools and drivers, jig saws, shrub and hedge trimmers, hammers, circular saws, grinders and polishers, finishing sanders, and edgers.

Many of Black & Decker's U.S. competitors that had already introduced products with double insulation had incurred a 15 to 20 percent premium in material and labor costs in doing so. It was Black & Decker's goal to add double insulation without increasing the cost of any new tool beyond that of the existing product. In addition, of course, Black & Decker aimed to avoid dilution of assets or return on investment.

In this instance, Black & Decker's decision to introduce fundamental redesign throughout its product line was motivated by the prospect of an industrywide requirement to incorporate double insulation in power tools. At other times, competitive product analysis plays an important role in decisions to redesign (the Appendix to this chapter describes a competitive product analysis carried out by the Sunbeam Appliance Company).

An important part of the plan for Double Insulation was the decision that the resources of the organization would be concentrated on this transition. Black & Decker would leave only a small portion of its management and engineering staff to carry out development efforts on new products. The development of new products was put in abeyance, and the resources usually devoted to development were focused on the manufacturing processes essential to the program.

To accomplish the engineering goals, a bridge was needed between design engineering and manufacturing. That bridge was the placement of advanced manufacturing engineers in residence at headquarters to work elbow to elbow with the design engineering groups. These manufacturing

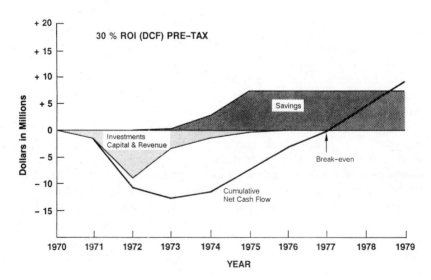

FIGURE 1   Financial analysis of Double Insulation program.

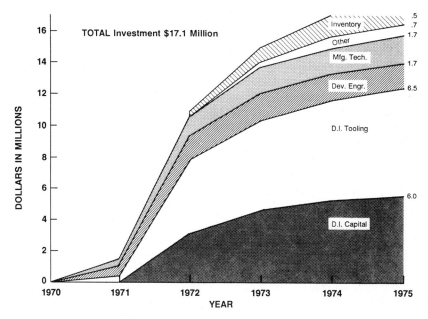

FIGURE 2   Investment requirements for Double Insulation program.

engineers were involved with machine development, process development, value and cost engineering, purchasing engineering, and packaging. In addition to bringing manufacturing and design together at the engineering level, the basic structure of the company was changed. Before these changes were made, the program structure had consisted of a general manager and two vice presidents—one for manufacturing, and one for engineering and product development. That organization was changed, and a new position—vice president of operations—was developed to combine manufacturing, product development, and advanced manufacturing engineering under one manager.

A final general point about the Double Insulation program was the large investment required and the long time horizon needed to reap a return on that investment. As Figure 1 shows, the break-even point in the program came nearly 7 years after the program began, and total cost was $17 million. Figure 2 shows the cumulative cost of the program from 1971 through 1975. Capital expenditures were $6 million. Tooling was $6.5 million, and development engineering and manufacturing technology were $1.7 million each. It is important to note that this program is rare from the standpoint that as much money was spent on manufacturing technology as on development engineering.

Designing for Manufacture

The transition to Black & Decker's leadership in cost and value was the result of collaborative effort among design, manufacturing, and manufacturing engineering functions. The changes in design and production of motors are one example of this collaborative effort.

The most common component in all power tools is the universal motor. Figure 3 shows all the components of such motors before redesign. Figure 4 shows the motor configuration both before and after redesign. Motors are now manufactured automatically, untouched by human hands. The laminations, placed at the head of the mechanized line, are stacked, welded, insulated, wound, varnished, terminated, and tested automatically. Table 2 shows, at 1974 volumes of 2,400 pieces per hour, that the new Double Insulation manufacturing system required 16 operators and that the old design would have required 108 operators. Material, labor, and overhead cost $0.51 in the old system and $0.31 in the new. The labor content cost $0.02 in the new system, down from $0.14 in the old.

Through attention to standardization, the entire range of Black & Decker power tools could be produced using a line of motors that vary only in stack length—that is, standardization froze the dimensional geometry of

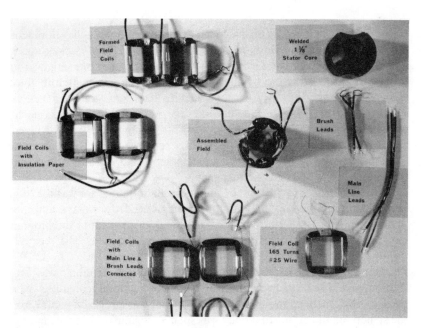

FIGURE 3   Electric motor field components.

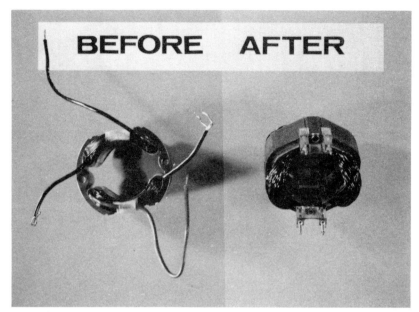

FIGURE 4   Motor configuration before and after redesign.

the motors in the axial profile. All motors can now be produced on the same machines, and the only difference is stack length and the amount of copper and steel used. As Figure 5 shows, designs ranged from 60 watts to 650 watts, and the only dimension that changed was in the axial profile. The only difference in cost from the low-wattage to the high-wattage motors was the cost of materials and machine time; labor cost remained the same through the entire wattage range.

TABLE 2   Motor Field Production, Operator Requirements, and Costs at 2,400 Units per Hour, Old and New Design and Manufacturing Process

	Old Design and Manufacturing Process	New Design and Manufacturing Process
Operators to produce	108	16
Cost to insulate (materials, labor, overhead)	$0.51	$0.31
Labor cost	$0.14	$0.02
Capital to produce	$400,000	$1,222,000
Annual savings (labor and materials only): $1,280,000		

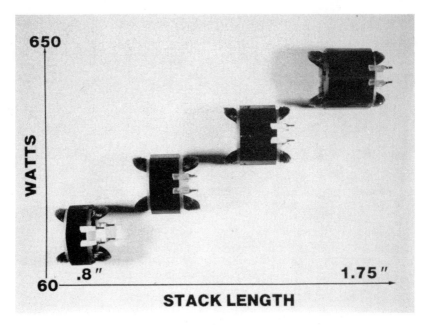

FIGURE 5   Motor stack length, 60 to 650 watts.

Another effect of design for manufacture can be seen in the production costs for the armature of the motors. As Table 3 shows, four times as many operators would have been needed to produce armatures under the old system as under the new system at a constant production volume. The effect on labor costs was dramatic. The labor cost of the manufacture using the new design was only $0.025 cents per unit, whereas the cost using the old design was $0.108 per unit.

TABLE 3   Armature Production, Operation Requirements, and Costs at 1,800 Units per Hour, Old and New Design and Manufacturing Processes

	Old Design and Manufacturing Process	New Design and Manufacturing Process
Operators to produce	60	15
Cost to insulate (materials, labor, overhead)	$0.26	$0.11
Labor cost	$0.108	$0.025
Capital to produce	$2,340,000	$795,000
Annual savings (labor and materials only): $540,000		

The Results of Double Insulation

The Double Insulation program worked for Black & Decker. It reduced production costs, created opportunities for profitable vertical integration, increased market share, and improved the company's capability for new product development. Each of these changes is discussed further in the following sections.

Cost Reductions

Cost reductions due to the Double Insulation program came mostly from labor savings, and the balance came from reduced factory overhead, material savings, and savings from standardization of parts and modularization. In 1976 Black & Decker reviewed the program and compared existing equipment and labor costs with the capital equipment and labor costs that would have been required for the 1976 volume without the Double Insulation program (see Table 4). If the company had not carried out this program, estimated 1976 requirements for motor manufacture would have been nearly 600 people whereas the new system required only 171. That is a labor cost difference—from $6.4 million down to $1.8 million—of $4.6 million per year. The capital investment for the new system was higher than simple capital replacement—$4.6 million instead of $3.0 million—but with labor savings of $4.6 million per year, the payback on the investment was 4 months.

In its 1974 annual report, Black & Decker published its assessment of

TABLE 4   Comparison of Labor and Capital Requirements for Electric Motor Production, 1972 and 1976 Volumes, Old and New Designs

	1972 Volume	1976 Volume
Workers required		
Old design	242	596
New design	86	171
Annual labor cost		
Old design	$2,700,000	$6,400,000
New design	$ 900,000	$1,800,000
Annual labor savings	$1,800,000	$4,600,000
Capital cost[a]		
Old design	$1,300,000	$3,000,000
New design	$2,300,000	$4,600,000
Capital cost difference	$1,000,000	$1,600,000
Payback	1.25 years	4 months

[a]Includes floor space at $20/sq. ft.

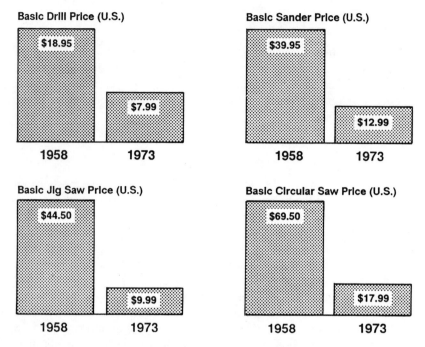

FIGURE 6   Prices, 1958 and 1973, of four basic hand power tools.

the effect of this project on four basic power tools (Figure 6). In current dollars, Black & Decker's power drills, for example, were $10 cheaper in 1973 than they were in 1958.

Figure 7 shows substantial reductions in the real cost of Black & Decker's products. The constant-dollar cost of products A, B, and C dropped by 47, 62, and 55 percent, respectively. The company was able to produce each product at an almost constant current dollar cost despite steady inflation in materials and labor costs. For Black & Decker's pricing position in the marketplace, the relevant comparison is between the top two lines on each graph, which show the difference, in current dollars, between manufacturing costs with and without Double Insulation.

Increased Vertical Integration

The cost and value leadership permitted unprecedented low prices to the consumers and thereby expanded Black & Decker's market share and increased household penetrations of power tools. The expanded volume resulted in opportunities for cost-effective vertical integrations. Examples include:

- Use of plastic materials grew from thousands of pounds per year to millions of pounds per year. Black & Decker's molding facilities were able to justify railcar bulk shipment of uncolored plastics resulting in a cost advantage of 5, 10, and sometimes 15 percent per pound. The coloring of plastic compounds at the molding machine reduced inventories, provided instant response to color changes, and eliminated material handling.
- The standardization of gears, and design revisions that allowed the change to spur gears from bevel gears, permitted the use of gears made from powdered metal. This change eliminated the need for gear cutting and hobbing, heat treating, and gauging. These activities all contributed to high capital cost, high labor cost, and inefficient use of material in production. The volumes were large enough to permit vertical integration of fabrication of powdered metal gears.
- Before the Double Insulation program, 29 percent of the total cost of a drill was in the cost of a purchased chuck. Production volumes, again, coupled with a modern state-of-art processing system enabled backward integration into chuck manufacturing at reduced costs of about 40 to 50 percent.
- Standardization of bearings, switches, cord sets, cartons, fasteners, and so on resulted in component volumes high enough to justify seeking sources on world markets for the best price.

The "inflation offset" idea proved to be recursive in that low production cost permitted low sales price, which increased unit sales, fueled vertical integration, and further reduced costs.

Competitive Performance and Market Share

In the U.S. market, Black & Decker's competitors in consumer power tools were caught absolutely flat-footed. Their product designs and manufacturing processes were costly, and in an attempt to continue to compete they tried to match Black & Decker prices. This diluted their profitability and collapsed their ability to redesign to match Black & Decker. In the resulting shakeout in the market for consumer power tools, Stanley, Skil, Pet, McGraw Edison, Sunbeam, General Electric, Wen, Thor, Porter Cable, and Rockwell all left the consumer market. Sears Roebuck and Co. was able to stay in the domestic consumer market with Black & Decker.

In the European market, consumer power tools were much more expensive because the tool offerings were different. European tool producers provided a power driver in the drill configuration, but all other power tools—sander, circular saw, hedge trimmer—were sold as attachments. The availability of new low-cost single-purpose power tools enabled the

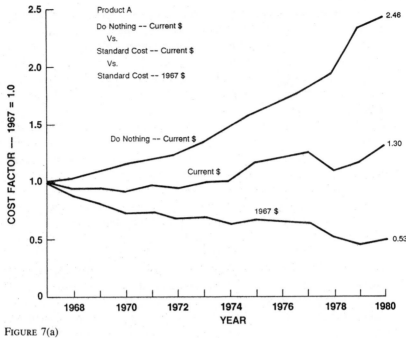

FIGURE 7(a)

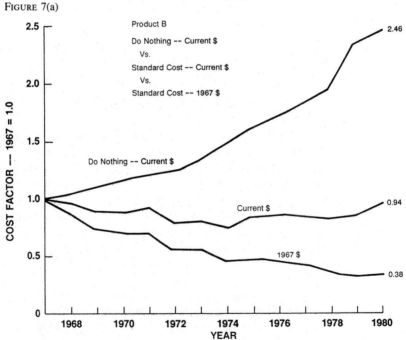

FIGURE 7(b)

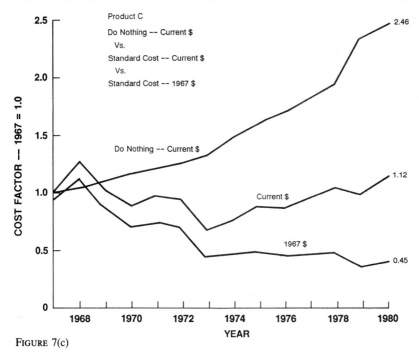

FIGURE 7(c)

FIGURE 7 Product cost trends in current dollars with and without Double Insulation and product cost trend with Double Insulation in constant 1967 dollars for three products, 1967-1980.

consumer to eliminate the inconvenience and performance compromise of attachment technology. Black & Decker performed well throughout Europe as the new tools both greatly expanded Black & Decker's market share and increased household penetration.

Impact on New Product Development

Another benefit of Black & Decker's efforts was a substantially improved ability in product development. As new product concepts emerged, much of the work in design and tooling was eliminated because of the standardization of motors, bearings, switches, gears, cord sets, and fasteners. Design and tooling engineers working on a new product had only to concern themselves with the "business end" of the product and to perfect its intended function. New designs could be developed using components already standardized for manufacturability. The product did not have to start with a blank sheet of paper and be designed from scratch.

As products reached their maturity and had to be dropped, massive write-offs and scrapping of tools and equipment were avoided because there were few special-purpose tools or equipment. This flexibility allowed marketers and managers to pivot quickly and avoid being tied to a dying product because they could not afford the write-offs.

In short, the pace of new product development and product retirement was greatly accelerated. Products could be introduced, exploited, and phased out with minimal expense related specifically to the decision to develop or retire a product.

SUMMARY AND CONCLUSIONS

In accomplishing the dramatic cost reductions through the Double Insulation program, the attitudes of the management were extremely important. Black & Decker management had a target of 15 percent compounded growth rate, and they wanted to remain independent and to service world markets. To do these things, the management focused not on marketing or financial manipulation, but on cost and value leadership in the industry. The management, as a team, projected how a successful program of this type could affect the marketplace and had the courage and tenacity to see it through. Black & Decker was also fortunate in having a large amount of latent talent: Many ordinary people proved capable of performing extraordinary tasks. With pricing and promotional strategies, the corporation was able to provide enough growth with the product improvements to avoid either reductions or expansions of its labor force.

Black & Decker's experience with the Double Insulation program shows the potential benefit of aggressively evaluating the design and production of "mature" manufactured products. Considered in relation to developed global markets, advances in materials and manufacturing processes provide opportunities for redesigning products to decrease the cost and increase the quality of manufactured goods.

Although the success that Black & Decker had with power tools may not be replicated in other industries, the principle of redesign and retooling for cost and value leadership in global markets can provide a focus for other manufacturing firms. It is a valid approach to achieving global competitiveness in manufacturing. By this means, U.S. firms can design, develop, and manufacture world-class products in the United States and at the same time achieve leadership in product value.

APPENDIX
Competitor Analysis by Sunbeam Appliance Company

In 1982 Sunbeam Appliance Company launched a program aimed at capturing at least 30 percent of the worldwide market share for the steam/dry iron in markets they wished to participate in. The first step was to assess global production capabilities and methods. Sunbeam obtained samples of competitive products from around the world for analysis of materials and labor content—estimated in time, not dollars. Components were reviewed and estimates of production costs were developed for all of the designs. After this material was pulled together, project management convened a 2-day review for Sunbeam engineers from Australia, Germany, England, Canada, Mexico, and the United States to talk over what had been uncovered in this global product evaluation.

That analysis revealed some interesting aspects of the design and manufacture of steam/dry irons around the world. The number of parts used in the product ranged from a high of 147 parts to a low of 74 parts. The number of fasteners ranged from 30 to 16, and the number of fastener types in any one design ranged from 15 to 9. Sunbeam's existing product used 97 parts with 18 fasteners in 10 configurations. Reducing the cost of that design, incorporating everything learned from composite design, yielded a design which had 73 parts, 13 fasteners, 7 types.

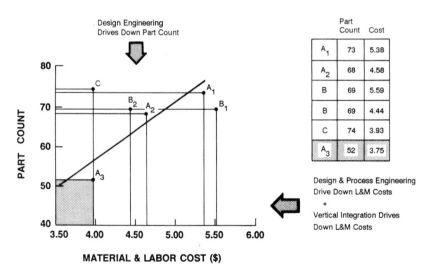

FIGURE A–1   Relationship of part count to material and labor cost per iron.

To gain a significant share of the market, however, it was necessary to leapfrog existing products and come up with a design with significantly lower cost and complexity over competitive offerings. Such a design was developed, with 51 parts and 3 fasteners in 2 configurations. Figure A-1 makes the point that driving down the part count also drives down cost.

Although reducing the part count entails a considerable effort in design engineering, effective design and process engineering will drive down labor and material costs.

The result of that effort was a composite design that would be the best of all of the products collected with attention to what the product would cost if the design were used throughout the world and compatibility were maintained. The new design is substantially cheaper to produce than either of Sunbeam's existing designs. The product was launched in 1986.

Capturing Value from Technological Innovation: Integration, Strategic Partnering, and Licensing Decisions

DAVID J. TEECE

Why, and under what circumstances, is the recognized technological progressiveness of a nation not sufficient to capture the benefits stemming from its capabilities in science and technology? This chapter examines why firms and nations can lose ground in the commercialization of advanced technologies at a time when they are the principal sources for major technological innovations of industrial significance; the capacity for scientific and technological innovation may be the last rather than the first advantage that a mature economy loses as it enters its declining phase.

The framework developed here helps identify the factors that determine who wins from innovation: The firm that is first to market, follower firms, or firms that have related capabilities that the innovator needs. The follower firms may or may not be imitators in the narrow sense of the term, although they sometimes are. The framework helps to explain the share of the profits from innovation accruing to the innovating firms and nations compared to its followers and suppliers.

THE PHENOMENON

A classic example of the phenomenon considered in this chapter is the computerized axial tomographic (CAT) scanner developed by the U.K. firm Electrical Musical Industries (EMI) Ltd.[1] By the early 1970s, EMI

This chapter is a revised version of a previously published paper by David J. Teece: "Profiting from Technological Innovation," *Research Policy*, Vol. 15(1986), No. 6.

was in a variety of product lines, including phonograph records, movies, and advanced electronics. EMI had developed high-resolution televisions in the 1930s, pioneered airborne radar during World War II, and developed the United Kingdom's first all solid-state computers in 1952.

In the late 1960s, the pattern recognition research of Godfrey N. Hounsfield, an EMI senior research engineer, resulted in his being able to display a scan of a pig's brain. Subsequent clinical work established that computerized axial tomography was viable for generating cross-sectional "views" of the human body, the greatest advance in radiology since the discovery of x rays in 1895.

Although EMI was initially successful with its CAT scanner, within 6 years of its introduction into the United States in 1973, the company had lost market leadership and by the eighth year had dropped out of the CAT scanner business. Other companies successfully dominated the market, though they were late entrants, and are still profiting in the business today.

A further example is that of the Royal Crown Companies, Inc., a small beverage company that was the first to introduce cola in a can and the first to introduce diet cola. Both Coca-Cola and Pepsi-Cola followed almost immediately and deprived Royal Crown of any significant advantage from its innovation. Bowmar Instrument Corporation, which introduced the

	INNOVATOR	IMITATOR–FOLLOWER
WIN	**1** Pilkington (Float Glass) G.D. Searle (NutraSweet) Dupont (Teflon)	**2** IBM (PC) Matsushita (VHS video recorders) Seiko (quartz watch)
LOSE	**4** RC Cola (diet cola) EMI (scanner) Bowmar (calculator) Xerox ("Star") DeHavilland (Comet)	**3** Kodak (instant photography) Northrup (F20) DEC (PC)

FIGURE 1 Taxonomy of outcomes from the innovation process.

pocket calculator, was not able to withstand competition from Texas Instruments, Hewlett-Packard, and others, and went out of business. Xerox Corporation failed to succeed with its entry into the office computer business, even though Apple Computer, Inc., succeeded with the MacIntosh, which contained many of Xerox's key product ideas, such as the mouse and icons. The story of the DeHavilland Comet has some of the same features. The Comet I jet was introduced into the commercial airline business 2 years or so before Boeing introduced the 707, but DeHavilland failed to capitalize on its substantial early advantage. MITS introduced the first personal computer, the Altair, experienced a burst of sales, then slid quietly into oblivion.

If there are innovators who lose, there must be followers (imitators) who win. A classic example is IBM Corporation with its PC, a great success from the time it was introduced in 1981. Neither the architecture nor the components of the IBM PC were considered advanced when introduced; nor was the way the technology was packaged a significant departure from the then-current practice. Yet the IBM PC was fabulously successful and established MS-DOS as the leading operating system for 16-bit PCs. By the end of 1984, IBM had shipped more than 500,000 PCs and may have irreversibly eclipsed Apple in the PC industry.

Figure 1 presents a simplified taxonomy—with examples—of the possible outcomes from innovation. Quadrant 1 represents positive outcomes for the innovator. A first-to-market advantage is translated into a sustained competitive advantage that either creates a new earnings stream or enhances an existing one. Quadrant 4 and its corollary quadrant 2 are the focus of this paper.

PROFITING FROM INNOVATION: BASIC BUILDING BLOCKS

To develop a coherent framework within which to explain the distribution of outcomes illustrated in Figure 1, three fundamental building blocks must be put in place: the appropriability regime, the dominant design paradigm, and complementary assets.

Regimes of Appropriability

A regime of appropriability refers to the environmental factors, excluding firm and market structure, that govern an innovator's ability to capture the profits generated by an innovation. The most important dimensions of such a regime are the nature of the technology and the efficacy of legal mechanisms of protection.

It has long been known that patents do not work in practice as they do in theory. Rarely, if ever, do patents confer perfect appropriability, although they do afford considerable protection on new chemical products and rather simple mechanical inventions. Many patents can be "invented around" at modest costs. They are especially ineffective at protecting process innovations. Often patents provide little protection, because the legal requirements for upholding their validity or for proving their infringement are high.

In some industries, particularly where the innovation is embedded in processes, trade secrets are a viable alternative to patents. Protection of trade secrets is possible, however, only if a firm can put its product before the public and still keep the underlying technology secret. Usually only chemical formulas and industrial-commercial processes (for example, cosmetics and recipes) can be protected as trade secrets after they are placed on the market.

The degree to which knowledge is tacit or codified also affects ease of imitation. Codified knowledge is easier to transmit and receive and is therefore more exposed to industrial espionage and the like. Tacit knowledge by definition is not articulated, and transfer is hard unless those who possess the know-how in question can demonstrate it to others (Teece, 1981). Survey research indicates that methods of appropriability vary markedly across industries, and probably within industries as well (Levin et al., 1984).

The property rights environment within which a firm operates can thus be classified according to the nature of the technology and the efficacy of the legal system to assign and protect intellectual property. Though a gross simplification, a dichotomy can be drawn between environments in which the appropriability regime is "tight" (technology is relatively easy to protect) and "loose" (technology is almost impossible to protect). Examples of the former include the formula for Coca-Cola syrup; an example of the latter is the Simplex algorithm in linear programming.

The Dominant Design Paradigm

Two stages are commonly recognized in the evolutionary development of a given branch of a science: the pre-paradigmatic stage when there is no single, generally accepted conceptual treatment of the phenomenon in a field of study, and the paradigmatic stage, which begins when a body of theory appears to have passed the canons of scientific acceptability. The emergence of a dominant paradigm signals scientific maturity and the acceptance of agreed-upon "standards" by which what has been referred to as "normal" scientific research can proceed. These "standards" remain in force unless the paradigm is overturned. Revolutionary science is what overturns normal science, as when the Copernican theories of astronomy overturned those of Ptolemy in the seventeenth century.

Abernathy and Utterback (1978), Dosi (1982), and Utterback (in this volume) provide treatment of the technological evolution of an industry in ways that parallel Kuhnian notions of scientific evolution (Kuhn, 1970). In the early stages of industrial development, product designs are fluid, manufacturing processes are loosely and adaptively organized, and generalized capital is used in production. Competition among firms manifests itself in competition among designs, which are markedly different from each other. This might be called the pre-paradigmatic stage of an industry.

After considerable trial and error in the marketplace, one design or a narrow class of designs begins to emerge as the most promising. Such design must be able to meet a set of user needs in a relatively complete fashion. The Model T Ford, the IBM System/360, and the Douglas DC-3 are examples of dominant designs in the automobile, computer, and aircraft industries, respectively.

Once a dominant design emerges, competition shifts to price and away from design. Competitive success then shifts to a new set of variables. Scale and learning become much more important, and specialized capital is deployed as incumbents seek to lower unit costs through exploiting economies of scale and learning. Reduced uncertainty over product design provides an opportunity to amortize specialized long-lived investments.

Innovation is not necessarily halted once the dominant design emerges; as Clarke (1985) points out, it can occur at a lower level in the design heirarchy. For instance, a "v" cylinder configuration emerged in automobile engine blocks during the 1930s with the Ford V-8 engine. Niches were quickly found for it. Moreover, once the product design stabilizes, there is likely to be a surge of process innovation as producers attempt to lower production costs for the new product (see Figure 2).

The Abernathy-Utterback framework does not characterize all industries. It seems better suited to mass markets, in which consumer tastes are relatively homogeneous, than to small niche markets where the absence of scale and learning economies attaches a much lower penalty to multiple designs. For these niche markets, generalized equipment will be used in production.

The emergence of a dominant design is a watershed that holds great significance for the distribution of profits between innovator and follower. The innovator may have been responsible for the fundamental scientific breakthroughs as well as the basic design of the new product. However, if imitation is relatively easy, imitators may enter the fray, modifying the product in important ways, yet relying on the fundamental designs pioneered by the innovator. When the game of musical chairs stops and a

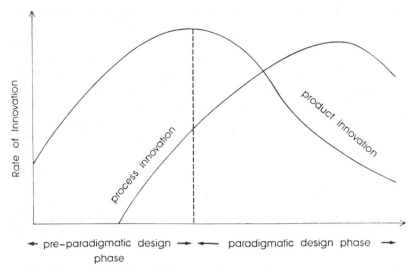

FIGURE 2   Innovation over the product/industry life cycle.

dominant design emerges, the innovator might well end up in a disadvantageous position relative to a follower. Hence, when imitation is coupled with design modification before the emergence of a dominant design, followers have a good chance that their modified product will be annointed as the industry standard, often to the great disadvantage of the innovator.

Complementary Assets

Let the unit of analysis be an innovation. An innovation consists of technical knowledge about how to do something better than the existing state of the art. Assume that the know-how in question is partly codified and partly tacit. For such know-how to generate profits, it must be sold or used in the market.

In almost all cases, the successful commercialization of an innovation requires that the know-how in question be used in conjunction with other capabilities or assets. Services such as marketing, competitive manufacturing, and after-sales support are almost always needed. These services are often obtained from complementary assets that are specialized. For example, the commercialization of a new drug is likely to require the dissemination of information over a specialized information channel. In some cases, as when the innovation is systemic, the complementary assets may be other parts of a system. For instance, computer hardware typically requires specialized software, both for the operating system and for applications. Even when an innovation is autonomous, as with plug-compatible components, certain complementary capabilities or assets will be needed for successful commercialization. Figure 3 summarizes this relationship schematically.

An important distinction is whether the assets required for least-cost production and distribution are specialized to the innovation. Figure 4 illustrates differences between complementary assets that are generic, specialized, and cospecialized.

Generic assets are general-purpose assets that need not be tailored to the innovation in question. Specialized assets are those where there is unilateral dependence between the innovation and the complementary asset. Cospecialized assets are those for which there is a bilateral dependence. For instance, specialized repair facilities were needed to support the introduction of the rotary engine by Mazda. These assets are cospecialized because of the mutual dependence of the innovation on the repair facility. Containerization similarly required the deployment of some cospecialized assets in ocean shipping and terminals. However, the dependence of truck-

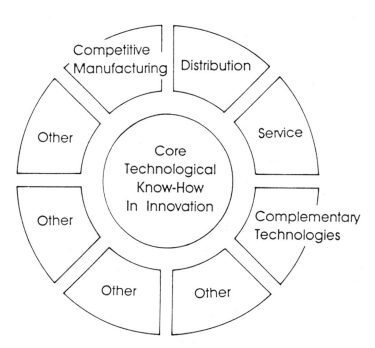

FIGURE 3   Complementary assets needed to commercialize an innovation.

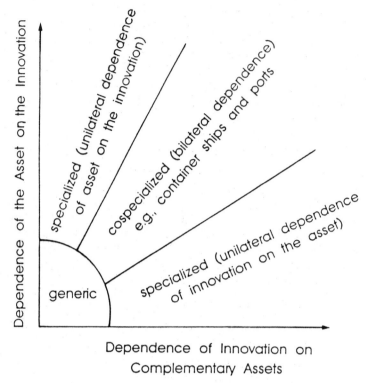

Dependence of Innovation on Complementary Assets

FIGURE 4 Complementary assets: generic, specialized, and cospecialized.

ing on containerized shipping was less than that of containerized shipping on trucking, as trucks can convert from containers to flatbeds at low cost. An example of a generic asset is the manufacturing facilities needed to make running shoes. Generalized equipment can be employed in the main, exceptions being the molds for the soles.

IMPLICATIONS FOR PROFITABILITY

These three concepts can now be related in a way that sheds light on the imitation process and the distribution of profits between innovator and follower. We begin by examining tight appropriability regimes.

Tight Appropriability Regimes

In those few instances when the innovator has an ironclad patent or copyright protection, or when the nature of the product is such that trade secrets effectively deny imitators access to the relevant knowledge, the innovator is almost assured that the innovation can be translated into market value for some period of time. Even if the innovator does not possess the desirable endowment of complementary costs, ironclad protection of intellectual property gives the innovator the time to acquire these assets. If these assets are generic, a contractual relationship may well suffice, and the innovator may simply license the technology. Specialized R&D firms are viable in such an environment. Universal Oil Products, an R&D firm that developed refining processes for the petroleum industry, is a case in point. If, however, the complementary assets are specialized or cospecialized, contractual relationships are exposed to hazards, because one or both parties will have to commit capital to certain irreversible investments, which will be valueless if the relationship between innovator and licensee breaks down. Accordingly, the innovator may find it prudent to expand by acquiring or developing specialized and cospecialized assets. Fortunately, the factors that render imitation difficult will enable the innovator to build or acquire those complementary assets without competing with imitators for their control.

Competition from imitators is muted in this type of regime, which sometimes characterizes the petrochemical industry. In this industry, the protection offered by patents is fairly easily enforced. One factor assisting the licensee in this regard is that most petrochemical processes are designed around a variety of catalysts that can be kept proprietary. An agreement not to analyze the catalyst can be extracted from licensees, affording extra protection. However, even if such requirements are violated by licensees, the innovator is still well positioned, as the most important properties of a catalyst are related to its physical structure, and the process for generating this structure cannot be deduced from structural analysis alone. Every chemical-reaction technology a company acquires is thus accompanied by an ongoing dependence on the innovating company for the catalyst appropriate to the plant design. Failure to comply with the licensing contract can thus result in a cutoff in the supply of the catalyst and possibly closure of the facility.

Similarly, if the innovator comes to market in the pre-paradigmatic phase with a sound product concept but the wrong design, a tight appropriability regime will afford the innovator the time needed to perform the trials needed to get the design right. As discussed earlier, the best initial design concepts often turn out to be hopelessly wrong, but if the innovator is protected by an impenetrable thicket of patents, or has technology that is difficult to copy, then the market may well afford the innovator the necessary time to develop the right design before being eclipsed by imitators.

Loose Appropriability

Tight appropriability is the exception rather than the rule. Therefore, innovators must turn to business strategy if they are to keep imitators at bay. The nature of the competitive process will depend on whether the industry is in the paradigmatic or pre-paradigmatic phase.

Pre-paradigmatic Phase

In the pre-paradigmatic phase, the innovator must be careful to let the basic design "float" until there is sufficient evidence that the design is likely to become the industry standard. In some industries there may be little opportunity for product modification. In microelectronics, for example, designs become locked in when the circuitry is chosen. Product modification is limited to debugging and modifying software. An innovator must begin the design process anew if the product does not fit the market well. In some respects, however, the selection of designs is dictated by the need to meet compatibility standards so that new hardware can be used with existing applications software. In one sense, therefore, the design issue for the microprocessor industry today is relatively straightforward: deliver greater power and speed while meeting the computer industry standards of the existing software base. However, from time to time windows of opportunity emerge for the introduction of entirely new families of microprocessors that will define a new industry and software standard. In these instances, basic design parameters are less well defined and can be permitted to "float" until market acceptance is apparent.

The early history of the automobile industry exemplifies the importance of selecting the right design in the pre-paradigmatic stages. None of the early producers of steam-powered cars survived the early shakeout when the internal combustion engine in a closed-body automobile emerged as the dominant design. The steamer, nevertheless, had numerous early virtues, such as reliability, which the cars with internal combustion engines could not deliver.

The British fiasco with the Comet I is also instructive. DeHavilland had picked an early design that had both technical and commercial flaws. By moving into production, significant irreversibilities and loss of reputation hobbled de Havilland to such a degree that it was unable to convert to the Boeing design that subsequently emerged as dominant. It was not even able to occupy second place, which went instead to Douglas.

As a general principle, it appears that innovators in loose appropriability regimes need to be intimately coupled to the market so that user needs can affect designs. When multiple parallel and sequential prototyping is feasible, it has clear advantages. Generally such an approach is prohibitively costly. When development costs for a large commercial aircraft exceed a billion dollars, variations on a theme are all that is possible.

Hence, assessing the probability that an innovator will enter the paradigmatic phase possessing the dominant design is problematic. The probabilities increase, the lower the relative cost of prototyping and the more tightly coupled the firm is to the market. The latter is a function of organizational design and can be influenced by managerial choices. The former is embedded in the technology and cannot be influenced, except in minor ways, by managerial decisions. Consequently, in industries with large costs for development and prototyping—hence significant irreversibilities—and where innovation of the product concept is easy, the probability that the innovator will emerge as a winner at the end of the pre-paradigmatic stage is low.

Paradigmatic Stage

In the pre-paradigmatic phase, complementary assets do not loom large. Rivalry focuses on trying to identify the design that will be dominant. Production volumes are low, and there is little to be gained in deploying specialized assets, as scale economies are unavailable and price is not a principal competitive factor. As the leading design or designs begin to be revealed by the market, however, volumes increase and opportunities for economies of scale induce firms to begin gearing up for mass production by acquiring specialized tooling and equipment and possibly specialized distribution as well. Since these investments impose significant irreversibilities, producers are likely to proceed with caution. Islands of specialized capital will begin to appear in an industry that otherwise features a sea of general-purpose manufacturing equipment.

But as the terms of competition begin to change, and prices become increasingly important, access to complementary assets becomes critical. Since the core technology is easy to imitate, by assumption, commercial success swings upon the terms and conditions affecting access to the required complementary assets.

It is at this point that specialized and cospecialized assets become critically important. Generalized equipment and skills, almost by definition, are always available in an industry, and even if they are unavailable, they do not entail significant irreversibilities. Accordingly, firms have easy access to this type of capital, and, even if the relevant assets are not available in sufficient quantity, they can easily be put in place as this involves few risks. Specialized assets, on the other hand, imply significant irreversibilities and cannot be easily acquired by contract, as the risks are significant for the party making the dedicated investment. The firms that control the cospecialized assets, such as distribution channels and specialized manufacturing capacity, are clearly in an advantageous position relative to an innovator. Indeed, in the rare instances in which incumbent firms possess an airtight monopoly over specialized assets, and the innovator is in a regime of loose appropriability, all of the profits from the innovation could conceivably accrue to those firms, who should be able to get the upper hand.

Even when the innovator does not face competitors or potential competitors who control key assets, the innovator may still be disadvantaged. For instance, the technology embedded in cardiac pacemakers was easy to imitate, so competitive outcomes quickly came to be determined by who had easiest access to the complementary assets—in this case, specialized marketing. A similar situation has recently arisen in the United States with respect to personal computers. As an industry participant recently observed:

> There are a huge number of computer manufacturers, companies that make peripherals (e.g., printers, hard disk drives, floppy disk drives), and software companies. They are all trying to get marketing distributors because they cannot afford to call on all of the U.S. companies directly. They need to go through retail distribution channels, such as Businessland, in order to reach the marketplace. The problem today, however, is that many of these companies are not able to get shelf space and thus are having a very difficult time marketing their products. The point of distribution is where the profit and the power are in the marketplace today. [Norman, 1986, p. 438]

CHANNEL STRATEGY ISSUES

The preceding analysis indicates how access to complementary assets, such as manufacturing and distribution, on competitive terms is critical if the innovator is to avoid handing over most of the profits to imitators, or to the owners of the complementary assets that are specialized or cospecialized to the innovation. It is now necessary to delve deeper into the control structure that the innovator ideally will establish over these critical assets.

There are many possible channels that could be employed. At one extreme the innovator could integrate into all of the necessary complementary assets, but complete integration is likely to be unnecessary and also prohibitively expensive. It is important to recognize that the variety of assets and competences needed is likely to be quite large, even for only modestly complex technologies. To produce a personal computer, for instance, a company needs access to expertise in semiconductor, display, disk drive, networking, and keyboard technologies, among others. No company can keep pace in all of these areas by itself. At the other extreme, the innovator could try to gain access to these assets through straightforward contractual relationships (for example, component supply contracts, fabrication contracts, service contracts). In many instances such contracts may suffice, although they sometimes expose the innovator to hazards and dependencies that might otherwise be avoided. Between the fully integrated and full contractual extremes are many intermediate forms and channels. An analysis of the properties of the two extreme forms is presented below. A brief synopsis of mixed modes then follows.

Contractual Modes

The advantages of a contractual solution—whereby the innovator signs a contract, such as a license, with independent suppliers, manufacturers, or distributors—are obvious. The innovator will not have to make the upfront capital expenditures needed to build or buy the assets in question. This reduces risks as well as cash requirements.

Contracting rather than integrating is likely to be the optimal strategy when the innovator's appropriability regime is tight and the complementary assets are available in competitive supply (that is, there is adequate capacity and a choice of sources).

Both conditions apply in the petrochemical industry, for instance, so an innovator does not need to be integrated to be successful. Consider, first, the appropriability regime. As discussed earlier, the protection offered by patents is fairly easily enforced, particularly for process technology, in the petrochemical industry. Given the advantageous feedstock prices available to hydrocarbon-rich petrochemical exporters, and the appropriability regime characteristic of this industry, there is neither incentive nor advantage in owning the complementary assets (production facilities), as they are not typically highly specialized to the innovation. Union Carbide appears to realize this and has recently adjusted its strategy accordingly. Essentially, Carbide is placing its existing technology into a new subsidiary, Engineering and Hydrocarbons Service. The company is engaging in licensing and offers engineering, construction, and management services to customers who want to take their feedstocks and integrate them forward into petrochemicals. But Carbide itself appears to be backing away from an integration strategy.

Chemical and petrochemical product innovations are not as easily protected as process technology is, which should raise new challenges to innovating firms in developed nations as they attempt to shift out of commodity petrochemicals. There are already numerous examples of new products that made it to the marketplace, filled a customer need, but never generated competitive returns to the innovator because of imitation. For example, in the 1960s the Dow Chemical Company decided to start manufacturing rigid polyurethene foam. It was quickly imitated, however, by many small firms that had lower costs.[2] The absence of low-cost manufacturing capability left Dow vulnerable.

Contractual relationships can bring added credibility to the innovator, especially if the innovator is relatively unknown when the contractual partner is established and viable. Indeed, arms-length contracting that embodies more than a simple buy-sell agreement is becoming so common, and is so multifaceted, that the term *strategic partnering* has been devised to describe it. Even large companies such as IBM are now engaging in it. For IBM, partnering buys access to new technologies, enabling a company to "learn things we couldn't have learned without many years of trial and error."[3] IBM's arrangement with Microsoft Corporation for the use of MS-DOS operating system software on the IBM PC facilitated the timely introduction of IBM's personal computer into the market.

Smaller, less integrated companies are often eager to sign on with established companies because of the name recognition and reputation spillovers. For instance, Cipher Data Products, Inc., contracted with IBM to develop a low-priced version of IBM's 3480 half-inch streaming cartridge drive, which is likely to become the industry standard. As Cipher management points out, "one of the biggest advantages to dealing with IBM is that, once you've created a product that meets the high quality standards necessary to sell into the IBM world, you can sell into any arena."[4] Similarly, IBM's contract with Microsoft "meant instant credibility" to Microsoft (McKenna, 1985, p. 94).

It is most important to recognize, however, that strategic (contractual) partnering, which is currently fashionable, holds certain hazards, particularly for the innovator, when the innovator is trying to use contracts to acquire specialized capabilities. First, it may be difficult to induce suppliers to make costly irreversible commitments that depend for their success on the success of the innovation. To expect suppliers, manufacturers, and distributors to do so is to invite them to take risks along with the innovator. The problem this poses for the innovator is similar to the problems associated with attracting venture capital. The innovator must persuade its prospective partner that the risk is a good one. The situation is open to opportunistic abuses on both sides. The innovator has incentives to overstate the value of the innovation, while the supplier has incentives to "run with the technology" should the innovation be a success.

Instances of irreversible capital commitments by both parties nevertheless exist. Apple's Laserwriter—a laser printer that allows PC users to produce near-typeset-quality text and art department graphics—is a case in point. Apple persuaded Canon, Inc., to participate in the development of the Laserwriter by providing subsystems from its copiers—but only after Apple contracted to pay for a certain number of copier engines and cases. In short, Apple accepted a good deal of the financial risk to induce Canon to assist in the development and production of the Laserwriter. The arrangement appears to have been prudent, yet there were clearly hazards for both sides. It is difficult to write, execute, and enforce complex development contracts, particularly when the design of the new product is still "floating." Apple was exposed to the risk that its coinnovator Canon would fail to deliver, and Canon was exposed to the risk that the Apple design and marketing effort would not succeed. Still, Apple's alternatives may have been limited, inasmuch as it did not command the requisite technology to "go it alone."

In short, the current euphoria over strategic partnering may be partially misplaced. The advantages are being stressed (for example, McKenna, 1985) without a balanced presentation of costs and risks. Briefly, there is the risk that the partner will not perform according to the innovator's perception of what the contract requires; there is the added danger that the partner may imitate the innovator's technology and attempt to compete with the innovator. This latter possibility is particularly acute if the provider of the complementary asset is uniquely situated with respect to the complementary asset in question and has the capacity to imitate the technology, which the innovator is unable to protect. The innovator will then find that it has created a competitor who is better positioned than the innovator to take advantage of the market opportunity at hand. *Business Week* has expressed concerns along these lines in its discussion of the "hollow corporation."[5]

It is important to bear in mind, however, that contractual or partnering strategies in certain cases are ideal. If the innovator's technology is well protected, and if what the partner has to provide is a "generic" capacity available from many potential partners, then the innovation will be able to maintain the upper hand while avoiding the costs of duplicating downstream capacity. Even if the partner fails to perform, adequate alternatives exist (by assumption, the partner's capacities are commonly available) so the innovator's efforts to successfully commercialize the technology ought to proceed profitably.

Integration Modes

Integration, which by definition involves ownership, is distinguished from pure contractual modes in that it typically facilitates incentive alignment and tighter organizational control (Williamson, 1985). An innovator who owns rather than rents the complementary assets needed to commercialize is in a position to capture spillover benefits stemming from increased demand for the complementary assets caused by the innovation.

Indeed, an innovator might be in the position, at least before the innovation is announced, to buy up capacity in the complementary assets, possibly to great subsequent advantage. If futures markets exist, though generally speaking they do not, taking forward positions in the complementary assets may suffice to capture much of the spillover.

Even after the innovation is announced, the innovator might still be able to build or buy complementary capacities at competitive prices if the innovation has ironclad legal protection (that is, if the innovation is in a tight appropriability regime). However, if the innovation is not tightly protected and once out is easy to imitate, then securing control of complementary capacities is likely to be the key success factor, particularly if those capacities are in fixed supply—so-called bottlenecks. Distribution and specialized manufacturing competences often become bottlenecks.

As a practical matter, however, an innovator may not have the time to acquire or build the complementary assets that ideally would be desirable. This is particularly true when imitation is easy, so that timing becomes critical. Additionally, the innovator may not have the financial resources to proceed. The implications of timing and cash constraints are summarized in Figure 5.

Accordingly, in loose appropriability regimes innovators need to rank complementary assets according to their importance. If the complementary assets are critical, ownership is warranted, although if the firm is cash constrained a minority position may well be a sensible approach.

When imitation is easy, strategic moves to build or buy specialized complementary assets must occur with due reference to the moves of competitors. There is no point in attempting to build a specialized asset, for instance, if one's imitators can do it faster and cheaper.

It should be self-evident that if the innovator is already a large enterprise with control over many of the relevant complementary assets, integration is not likely to be the issue it might otherwise be, as the innovating firm will already control many of the relevant specialized and cospecialized assets. However, in industries experiencing rapid technological change, it is unusual that a single company has the full range of expertise needed to bring advanced products to market in a timely and cost-effective way. Hence, the integration issue is of concern to both large and small firms.

Integration Versus Contract Strategies: An Analytic Summary

Figure 6 summarizes some of the relevant considerations in the form of a decision flow chart. It indicates that a profit-seeking innovator faced

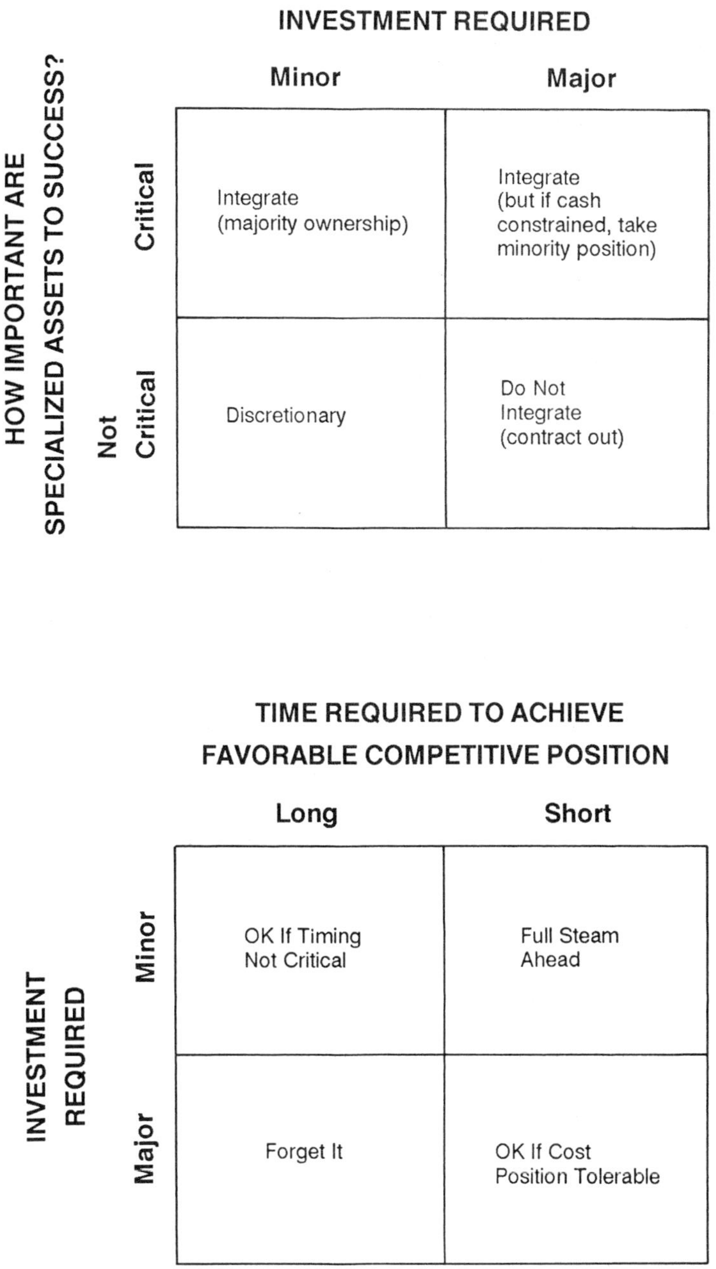

FIGURE 5   Specialized complementary assets and loose appropriability: integration calculus.

with weak protection of intellectual property and the need to access specialized complementary assets or capabilities is forced to expand through integration to prevail over imitators. Put differently, innovators who develop new products that possess poor protection of intellectual property but require specialized complementary capacities are more likely to parlay their technology into a commercial advantage rather than see it prevail in the hands of imitators.

Figure 6 makes it apparent that difficult strategic decisions arise when the appropriability regime is loose and when specialized assets are critical to profitable commercialization. These situations are common and require that a thorough assessment of competitors be part of the innovator's strategic assessment of opportunities and threats. Figure 7 carries this discussion a step further and considers only situations where commercialization requires certain specialized capabilities. It shows the appropriate strategies for the innovators and predicts the expected outcomes for the various players.

Three classes of players are of interest: innovators, imitators, and the owners of cospecialized assets (for example, distributors). All three can potentially benefit or lose from the innovation process. The latter can potentially benefit from the additional business that the innovation may direct in the asset owner's direction. Should the asset turn out to be a bottleneck with respect to commercializing the innovation, the owner of the bottleneck facilities is obviously in a position to extract profits from the innovator or the imitators.

The vertical axis in Figure 7 measures how those who possess the technology (the innovator or possibly the imitators) are positioned with respect to those firms that possess required specialized assets. The horizontal axis measures the "tightness" of the appropriability regime, tight regimes being evidenced by ironclad legal protection coupled with technology that is difficult to copy; loose regimes offer little in the way of legal protection, and the essence of the technology, once released, is transparent to the imitator. Loose regimes are further subdivided according to how the innovator and imitators are positioned in relation to each other. This is likely to be a function of factors such as lead time and prior positioning in the requisite complementary assets.

Figure 7 makes it apparent that even when firms pursue the optimal strategy, other industry participants may take the jackpot. This possibility is unlikely when the intellectual property in question is tightly protected. The only serious threat to the innovator is where a specialized complementary asset is "locked up," a possibility recognized in cell 4. This can rarely be done without the cooperation of government. But it frequently occurs, as when a foreign government closes access to a foreign market,

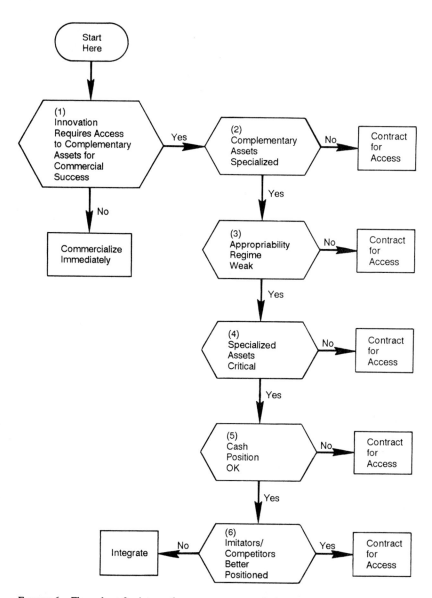

FIGURE 6 Flow chart for integration versus contract design.

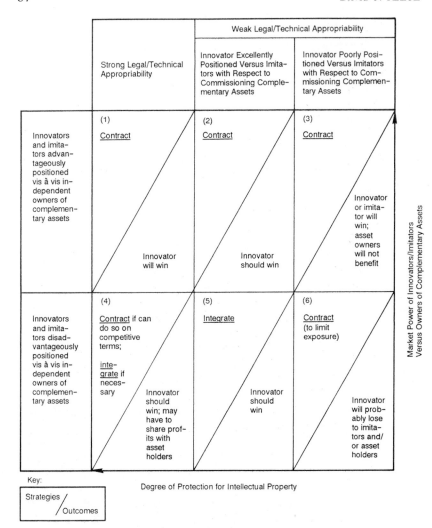

FIGURE 7   Optimal contract and integration strategies and outcomes for innovators: specialized asset case.

forcing the innovators to license to foreign firms, but with the government effectively cartelizing the potential licensees. With weak intellectual property protection, however, it is clear that the innovator will often lose out to imitators or asset holders, even when the innovator is pursuing the appropriate strategy (cell 6). Clearly, incorrect strategies can compound problems. For instance, if innovators integrate when they should contract, a heavy commitment of resources will be incurred for little if any strategic benefit, thereby exposing the innovator to even greater losses than would otherwise be the case. On the other hand, if an innovator tries to contract for the supply of a critical capability when it should build the capability itself, it may well find it has nurtured an imitator better able to serve the market.

Mixed Modes

The real world rarely provides extreme or pure cases. Decisions to integrate or license involve trade-offs, compromises, and mixed approaches. It is not surprising, therefore, that the real world is characterized by mixed modes of organization, involving judicious blends of integration and contracting. Sometimes mixed modes represent transitional phases. For instance, because of the convergence of computer and telecommunication technology, firms in each industry are discovering that they often lack the technical capabilities needed in the other. Since the technological interdependence of the two requires collaboration among those who design different parts of the system, intense cross-boundary coordination and information flows are needed. For separate enterprises, agreement must be reached on complex protocol issues among parties who see their interests differently. Contractual difficulties can be anticipated, as the selection of common technical protocols among the parties will often be followed by transaction-specific investments in hardware and software. There is little doubt that this was the motivation behind IBM's 1983 purchase of 15 percent of the Rolm Corporation, manufacturer of business communications systems. This position was expanded to 100 percent in 1984. IBM's stake in Intel Corporation, which began with a 12 percent purchase in 1982, is most probably not a transitional phase leading to 100 percent purchase, because both companies realized that the two corporate cultures are not compatible, and IBM may not be as impressed with Intel's technology as it once was.

The CAT Scanner and the IBM PC:
Insights from the Framework

EMI's failure to reap significant returns from the CAT scanner can be explained in large measure by reference to the concepts developed above. The scanner that EMI developed was of a technical sophistication much higher than would normally be found in a hospital, requiring a high level of training support and servicing. EMI did not possess these capabilities, could not easily contract for them, and was slow to realize their importance.

It most probably could have formed a partnership with a company like Siemens to gain access to the requisite capabilities. Its failure to do so was a strategic error compounded by the limited protection afforded by the law for the intellectual property embodied in the scanner. Although subsequent court decisions have upheld some of EMI's patent claims, once the product was in the market it could be reverse engineered and its essential features copied.

Two competitors, General Electric and Technicare, already possessed the complementary capabilities that the scanner required, and they were also technologically capable. In addition, both were experienced marketers of medical equipment and had reputations for quality, reliability, and service. GE and Technicare were thus able to commit their R&D resources to developing a competitive scanner and improving on the EMI scanner where they could while they rushed to market. GE began taking orders in 1976 and soon after made inroads on EMI's lead. In 1977 concern for rising health care costs caused the Carter administration to introduce "certificate of need" regulation, which required approval by the Department of Health, Education, and Welfare for expenditures on big ticket items like CAT scanners. This severely cut the size of the available market.

By 1978 EMI had lost the leadership in market share to Technicare, who was in turn quickly overtaken by GE. In October 1979 Geoffrey Hounsfield of EMI shared the Nobel Prize for invention of the CAT scanner. Despite this honor, and the public recognition of EMI's role in bringing this medical breakthrough to the world, the collapse of its scanner business forced EMI in the same year into the arms of a rescuer, Thorn Electrical Industries, Ltd. GE subsequently acquired what was EMI's scanner business from Thorn.[6] Though royalties continued to flow to EMI, the company had failed to capture the largest part of the profits generated by the innovation it had pioneered and successfully commercialized.

If EMI illustrates how a company with outstanding technology and an excellent product can fail to profit from innovation while the imitators succeed, the story of the IBM PC indicates how a new product representing only a modest technological advance can yield remarkable returns to the developer.

The IBM PC, introduced in 1981, succeeded despite the fact that its architecture was ordinary and its components standard. Philip Estridge's design team in Boca Raton, Florida, decided to use existing technology rather than the state of the art to produce a solid, reliable microcomputer. With a 1-year mandate to develop a PC, Estridge's team could do little else.

However, the IBM PC did use what at the time was a new 16-bit microprocessor (the Intel 8088) and a new disk operating system (DOS) adapted for IBM by Microsoft. Other than the microprocessor and the operating system, the IBM PC incorporated existing microcomputer "standards" and used off-the-shelf parts from outside vendors. IBM did write its own basic input/output system (BIOS), which is embedded in a read-only memory chip, but this was a relatively straightforward programming exercise.

The key to the PC's success was not the technology. It was the set of complementary assets that IBM either had or quickly assembled around the PC. To expand the market for PCs, there was a clear need for an expandable, flexible microcomputer system with extensive applications of software. IBM could have based its PC system on its own patented hardware and copyrighted software. Such an approach would cause complementary products to be cospecialized, forcing IBM to develop peripherals and a comprehensive library of software in a short time. Instead, IBM adopted what might be called an induced contractual approach. By adopting an open system architecture, as Apple had done, and by making technical information about the operating system publicly available, IBM induced a spectacular output of software by third-party suppliers. IBM estimated that by mid-1983, at least 3,000 hardware and software products were available for the PC.[7] Put differently, IBM pulled together the complementary assets, particularly software, required for success and did not even use contracts, let alone integration. This was despite the fact that the software developers were creating assets that were in part cospecialized to the IBM PC, at least in the first instance.

Several special considerations made this approach a reasonable risk for the software writers. A critical element was IBM's name and commitment to the project. The reputation behind the letters *I, B, M* is perhaps the greatest cospecialized asset the company possesses. The name implied that the product would be marketed and serviced in the IBM tradition. It guaranteed that PC-DOS would become an industry standard, so that the software business would not be solely dependent on IBM, because emulators were sure to enter. It guaranteed access to retail distribution outlets on competitive terms. The consequence was that IBM was able to take a product that was at best a modest technological accomplishment and turn it into a fabulous commercial success. The case demonstrates the role of complementary assets in determining outcomes.

Though the success of the IBM PC is ongoing, the appearance of machines compatible with the IBM PC (IBM compatibles and "clones") has somewhat attenuated PC market growth for IBM. The emergence and rapid acceptance of the IBM PC established a market-based software standard. Given IBM's reputation and the quality of the product, the emergence of a market standard was predictable, as was increased price competition as competitors focused on cost reduction and performance enhancement. The fact that IBM is no longer the overwhelmingly dominant PC manufacturer—possibly because of its price umbrella and modest rate of performance improvement—does not diminish the lesson of the IBM PC program with regard to capturing returns from innovation. Despite competition from compatibles and clones, IBM's return on investment must surely have been attractive.

IMPLICATIONS FOR R&D STRATEGY, INDUSTRY STRUCTURE, AND TRADE POLICY

Allocating R&D Resources

The analysis so far assumes that the firm has developed an innovation for which a market exists. It indicates the strategies that the firm must follow to maximize its share of industry profits relative to imitators and other competitors. There is no guarantee of success even if optimal strategies are followed.

The innovating firm can improve its total return to R&D, however, by adjusting its R&D investment portfolio to maximize the probability that the technological discoveries that emerge will be easy to protect with existing property law or will require for commercialization cospecialized assets already within the firm's repertoire of capabilities. Put differently, if an innovating firm does not target its R&D resources toward new products and processes that it can commercialize advantageously relative to potential imitators or followers, then it is unlikely to profit from its investment in R&D. In this sense, a firm's history—and the assets it already has in place—ought to condition its R&D investment decisions. Clearly, an innovating firm with considerable assets already in place is free to strike out in new directions, so long as it is aware of the kinds of capabilities required to commercialize the innovation. It is therefore clear that the R&D investment decision cannot be divorced from the strategic analysis of markets and industries, and the firm's position in them.

Small Firm Versus Large Firm Comparisons

Business commentators frequently remark that many small entrepreneurial firms that generate new, commercially valuable technology fail at the same time that a large multinational firm, even with a less meritorious record in innovation, will survive and prosper. One explanation of this phenomenon is now clear. Large firms are more likely to possess the relevant specialized and cospecialized assets at the time a new product is introduced. They can therefore do a better job of using their technology, however meager, to maximum advantage. Small domestic firms are less likely to have the relevant specialized and cospecialized assets within their boundaries. They must therefore incur the expense of trying either to build the necessary assets or to develop coalitions with competitors or with owners of the assets.

Regimes of Appropriability and Industry Structure

In industries where legal methods of protection are effective, or where new products are just hard to copy, the strategic necessity for innovating firms to obtain cospecialized assets would appear to be less compelling than in industries where legal protection is weak. In cases where legal protection is weak or nonexistent, the control of cospecialized assets will be needed for long-run survival.

In this regard, it is instructive to examine the U.S. drug industry (Temin, 1979). In the 1940s, the U.S. Patent Office began to grant patents on certain natural substances that involved difficult extraction procedures. Thus, in 1948 Merck received a patent on streptomycin, which is a natural substance. However, it was not the extraction process but the drug itself that received the patent. Hence, patents were important to the drug industry, but they did not prevent imitation as, in some cases, just changing one molecule would enable a company to come up with a similar substance not violating the patent (Temin, 1979, p. 436). Had patents been more inclusive—and this is not to suggest that they should be—licensing would have been an effective mechanism for Merck to profit from its innovation. The emergence of close substitutes for patented drugs, coupled with FDA regulation that had the effect of reducing the elasticity of demand for drugs, placed high rewards on a strategy of product differentiation. This strategy required extensive marketing, including a sales force that could directly contact doctors, who were the purchasers of drugs through their ability to create prescriptions.[8] The result was exclusive production (that is, the earlier industry practice of licensing was dropped) and forward integration into marketing (the relevant cospecialized asset).

Generally, if legal protection of the innovators' profits is secure, innovating firms can select their boundaries according to their ability to identify user needs and respond to those needs through research and development. The weaker the legal methods of protection, the greater the incentive to obtain the relevant cospecialized assets. Hence, as industries in which legal protection is weak begin to mature, integration into innovation-specific cospecialized assets will occur. Often this will take the form of backward, forward, and lateral integration. (Conglomerate integration is not part of this phenomenon.) For example, IBM's purchase of Rolm can be seen as a response to the impact of technological change on the identity of the cospecialized assets relevant to IBM's future growth.

Industry Maturity, New Entry, and History

As technologically progressive industries mature, and a greater proportion of the relevant cospecialized assets are brought under the corporate umbrellas of incumbents, new entry becomes more difficult. Moreover, when it does occur it is more likely to include the early formation of coalitions. Incumbents will own the cospecialized assets, and new entrants will find it necessary to forge links with them. Here lies the explanation for the sudden surge in strategic partnering now occurring internationally, and particularly in the computer and telecommunications industry. Note that this change should not be interpreted in anticompetitive terms. Given existing industry structure, coalitions ought to be seen not as attempts to stifle competition, but as mechanisms for lowering entry requirements for innovators.

In industries in which a technological change has occurred and required deployment of specialized or cospecialized assets, a configuration of firm boundaries that no longer have compelling efficiencies may well have arisen. Considerations that once dictated integration may no longer hold, yet there may not be strong forces leading to divestiture. Hence existing firm boundaries in some industries—especially those where the technological trajectory and attendant specialized asset requirements have changed—may be fragile. In short, history is important in understanding the structure of the modern business enterprise. Existing firm boundaries cannot always be assumed to have an obvious rationale in relation to today's requirements.

The Importance of Manufacturing to International Competitiveness

Practically all forms of technological know-how must be embedded in goods and services to yield value to the consumer. An important policy issue for the innovating nation is whether the identity of the firms and nations performing this function is important.

In a world of tight appropriability and zero transactions cost—the world of neoclassical theory—it is a matter of indifference whether an innovating firm has an in-house manufacturing capability, domestic or foreign. The firm can engage in arms-length contracting (patent licensing, know-how licensing, coproduction, and so on) for the sale of the output of the activity in which it has a comparative advantage (in this case R&D) and will maximize returns by specializing in what it does best.

However, in a regime of loose appropriability, and especially where the requisite manufacturing assets are specialized to the innovation, which is often the case, participation in manufacturing may be necessary if an innovator is to appropriate the rents from its innovation. Hence, if an innovator's manufacturing costs are higher than those of an imitator, the innovator may well be put in the position of ceding the largest share of profits to the imitator.

In a loose appropriability regime, low-cost imitator-manufacturers could capture all of the profits from innovation. In a loose appropriability regime where specialized manufacturing capabilities are necessary to produce new products, an innovator with a manufacturing disadvantage may find that an early advantage at the research and development stage has no commercial value. This is a potentially crippling situation unless the innovator is assisted by governmental processes. For example, one reason why U.S. manufacturers did not capture the greatest part of the profits from the development of color TV, for which RCA was primarily responsible, is that RCA and its U.S. licensees were not competitive at manufacturing. In this context, concern that the decline of manufacturing threatens the entire economy appears to be well founded.

A related implication is that as the technology gap closes, the basis of competition in an industry will shift to the cospecialized assets. This appears to be what is happening in microprocessors. Intel is no longer out ahead technologically. As Gordon Moore, CEO of Intel points out, "Take the top 10 [semiconductor] companies in the world . . . and it is hard to tell at any time who is ahead of whom. . . . It is clear that we have to be pretty damn close to the Japanese from a manufacturing standpoint to compete."[9] It is not just that strength in one area is necessary to compensate for weakness in another. As technology becomes more public and less proprietary through easier imitation, strength in manufacturing and other areas is necessary to benefit from whatever technological advantages an innovator may possess.

Put differently, the notion that the United States can adopt a "designer role" in international commerce while letting independent firms in countries such as Japan, Korea, Taiwan, or Mexico do the manufacturing is unlikely to be a successful strategy for the long run. This is because profits will accrue primarily to the low-cost manufacturers (by providing a larger sales base over which they can exploit their special skills). Where imitation is easy, and even where it is not, it is difficult to do business in the market for know-how (Teece, 1981). In particular, there are difficulties in pricing an intangible asset whose true performance features are difficult to predict.

The trend in international business toward what Miles and Snow (1986) call dynamic networks—characterized by vertical disintegration and contracting—therefore ought to be viewed with concern. Dynamic networks, or hollow corporations, may reflect innovative organizational forms not so much as the disassembly of the modern corporation because of deterioration in manufacturing and other activities that complement technological innovation. Dynamic networks may therefore signal not so much the rejuvenation of American enterprise as its piecemeal demise.

How Trade and Investment Barriers Affect Innovators' Profits

In regimes of loose appropriability, governments can move to shift the distribution of the gains from innovation away from foreign innovators and toward domestic firms by denying innovators ownership of specialized assets. The foreign firm, by assumption an innovator, will be left with the option of selling its intangible assets on the market for know-how if both trade and investment are foreclosed by government policy. This option may appear better than the alternative (no remuneration at all from the market in question). Licensing may then appear profitable, but only because access to the complementary assets is blocked by government.

Thus, when an innovating firm generating profits needs access to complementary assets abroad, host governments, by limiting access, can sometimes milk the innovators for a share of the profits, particularly that portion that originates from sales in the host country. However, the ability of host governments to do so depends on the importance of the host country's assets to the innovator. If the cost and infrastructure characteristics of the host country are such that it is the world's lowest cost manufacturing site, and if domestic industry is competitive, then by acting as a monopsonist the government of the host country ought to be able to adjust the terms of access to the complementary assets to appropriate a greater share of the profits generated by the innovation.[10] If, on the other hand, the host country offers no unique complementary assets except access to its own market, restrictive practices by the government will only redistribute profits with respect to domestic rather than worldwide sales.

Implications for the International Distribution of the Benefits from Innovation

Thus, it is clear that innovators who do not have access to the relevant specialized and cospecialized assets may end up ceding profits to imitators and other competitors, or to the owners of the specialized or cospecialized assets. Even when the innovator possesses the specialized assets, they may be located abroad. Foreign factors of production are thus likely to benefit from research and development occurring across borders. There is little doubt, for instance, that the inability of many U.S. multinationals to sustain competitive manufacturing in the United States results in declining returns to U.S. labor. Stockholders and top management probably do as well if not better when a multinational gains access to cospecialized assets in the firm's foreign subsidiaries. However, if there is unemployment in the factors of production supporting these assets, then the foreign factors of production will benefit from innovation originating beyond national borders. This shows how important it is that innovating nations maintain competence and competitiveness in the assets—especially manufacturing—that complement technological innovation. It also shows how important it is that innovating nations enhance the protection afforded worldwide to intellectual property.

It must be recognized, however, that there are inherent limits to the legal protection of intellectual property and that business and national strategies are therefore likely to be critical in determining how the gains from innovation are shared worldwide. By making the correct strategic decision, innovating firms can move to protect the interests of stockholders. But to ensure that domestic rather than foreign cospecialized assets capture the largest share of the externalities spilling over to complementary assets, the supporting infrastructure for those complementary assets must not be allowed to decay. In short, if a nation has prowess at innovation, then in the absence of ironclad protection for intellectual property, it must maintain well-developed complementary assets if it is to capture the spillover benefits from innovation.

CONCLUSION

This chapter has attempted to synthesize from recent research in industrial organization and strategic management a framework within which to analyze the distribution of the profits from innovation. The framework indicates that the boundaries of the firm are an important strategic variable for innovating firms. The ownership of complementary assets, particularly when they are specialized or cospecialized, helps establish who wins and who loses from innovation. Imitators can often outperform innovators if they are better positioned with respect to critical complementary assets. Hence, public policy aimed at promoting innovation must focus not only on R&D but also on complementary assets as well as the underlying infrastructure. If government decides to stimulate innovation, it is important to eliminate barriers to the development of complementary assets that are specialized or cospecialized to innovation. To fail to do so will cause a large portion of the profits from innovation to flow to imitators and other competitors. If these firms lie beyond one's national borders, there are obvious implications for the international distribution of income.

When applied to world markets, results similar to those obtained from the "new trade theory" are suggested by the framework. In particular, tariffs and other restrictions on trade can in some cases injure innovating firms while simultaneously benefiting protected firms when they are imitators. However, the propositions suggested by the framework vary according to appropriability regimes, suggesting that economywide conclusions will be elusive. The policy conclusions for commodity petrochemicals, for instance, are likely to differ from those for semiconductors.

The approach also suggests that the product life cycle model of international trade will play itself out differently in different industries and markets, in part according to appropriability regimes and the nature of the assets needed to convert a technological success into a commercial one. Whatever its limitations, the approach establishes that it is not so much the structure of markets as the structure of firms, particularly the scope of their boundaries, coupled with national policies on the development of complementary assets, that determines the distribution of the profits among innovators and imitator-followers.

ACKNOWLEDGMENTS

I wish to thank Raphael Amit, Harvey Brooks, Chris Chapin, Therese Flaherty, Richard Gilbert, Bruce Guile, Heather Haveman, Mel Horwitch, David Hulbert, Carl Jacobson, Michael Porter, Gary Pisano, Richard Rumelt, Richard Nelson, Raymond Vernon, and Sidney Winter for helpful discussions relating to the subject matter of this paper. I gratefully acknowledge the financial support of the National Science Foundation under grant no. SRS-8410556 to the Center for Research in Management, University of California, Berkeley. Versions of this paper were presented at a National Academy of Engineering symposium titled "World Technologies and National Sovereignty," February 1986; at a conference on innovation at the University of Venice, March 1986; and at seminars at the Massachusetts Institute of Technology and Harvard, Yale, and Stanford universities. Helpful comments received at these conferences and seminars are gratefully acknowledged.

NOTES

1. The EMI story is summarized in Michael Martin, *Managing Technological Innovation and Entrepreneurship* (Reston, Va.: Reston Publishing Company, 1984).

2. Executive Vice President, Union Carbide, Robert D. Kennedy, quoted in *Chemical Week*.
3. Comment attributed to Peter Olson III, IBM's director of business development, as reported in "The Strategy Behind IBM's Strategic Alliances," *Electronic Business*, October 1, 1985, p. 126.
4. Comment attributed to Norman Farquhar, Cipher's vice president for strategic development, as reported in *Electronic Business*, October 1, 1985, p. 128.
5. See *Business Week*, March 3, 1986, pp. 57-59. *Business Week* uses the term "hollow corporation" to describe a firm that lacks in-house manufacturing capability.
6. See "GE Gobbles a Rival in CT Scanners," *Business Week*, May 19, 1980.
7. F. Gens and C. Christiansen, "Could 1,000,000 IBM PC Users Be Wrong," *Byte*, November 1983, p. 88.
8. In the period before FDA regulation, all drugs other than narcotics were available without prescriptions. Since the user could purchase drugs directly, sales were price sensitive. Once prescriptions were required, this price sensitivity collapsed; doctors not only do not have to pay for the drugs, but in most cases they are unaware of the prices of the drugs they are prescribing.
9. "Institutionalizing the Revolution," *Forbes*, June 16, 1986, p. 35.
10. If the host country market structure is monopolistic, private actors might be able to achieve the same benefit. Government can force collusion of domestic enterprises to their mutual benefit.

REFERENCES

Abernathy, W. J., and J. M. Utterback. 1978. Patterns of industrial innovation. Technology Review 80:7(June/July):40-47.
Clarke, K. B. 1985. The interaction of design hierarchies and market concepts in technological evolution. Research Policy 14:235-251.
Dosi, G. 1982. Technological paradigms and technological trajectories. Research Policy 11(3):147-162.
Kuhn, T. 1970. The Structure of Scientific Revolutions, 2nd ed. Chicago: University of Chicago Press.
Levin, R., A. Klevorick, N. Nelson, and S. Winter. 1984. Survey research on R&D appropriability and technological opportunity. Yale University. Unpublished manuscript.
McKenna, R. 1985. Market positioning in high technology. California Management Review XXVII:3(Spring):82-108
Miles, R. E., and C. C. Snow. 1986. Network organizations: New concepts for new forms. California Management Review XXVIII:3(Spring):62-73.
Norman, D. A. 1986. Impact of entrepreneurship and innovations on the distribution of personal computers. Pp.437-439 in R. Landau and N. Rosenberg, eds., The Positive Sum Strategy. Washington, D.C.: National Academy Press.
Teece, D. J. 1981. The market for know how and the efficient international transfer of technology. Annals of the American Academy of Political and Social Science 458(November):81-96
Temin, P. 1979. Technology, regulation, and market structure in the modern pharmaceutical industry. The Bell Journal of Economics 10(2):429-446.
Williamson, O. E. 1985. Economic Institutions of Capitalism. New York: The Free Press.

International Industries:
Fragmentation Versus Globalization

YVES DOZ

Since World War II, growth in international trade has exceeded world economic growth by a substantial margin, and national economies have become increasingly dependent on world trade. Up to 50 percent of the gross national product (GNP) of small European countries is traded internationally, whereas only about 25 percent of GNP in larger European countries and 10 to 15 percent of GNP in the comparatively isolated large economies of the United States and Japan is traded internationally. Markets for many industrial goods have become increasingly homogeneous. Simultaneously, foreign investment has grown rapidly, both in developed and in developing countries.[1] Not only has the total stock of capital grown rapidly, but, more significantly, there has been growth in the number of subsidiaries of multinational companies (MNCs); growth in the number of countries in which specific firms were active; and increasing diversity in the products manufactured and sold abroad through subsidiaries of MNCs (Vernon and Davidson, 1979).

As both international trade and investment grew rapidly, international competition became more intense, and many national industries became global industries. Similarity of markets in different countries and intense global competition drove international competitors to coordinate their market and competitive strategies between countries more actively. The relevant scope of strategy thus shifted from discrete national markets to global markets, and coordinated worldwide competitive actions between the various subsidiaries of MNCs became more important.

As national competition shifted to global competition, foreign investment also shifted. Protectionism in the 1930s, the trauma of World War II, and national reconstruction policies led the early multinational investors to fragment their operations into discrete market-servicing, self-sufficient investments with little interdependence between operations in separate countries. The developing countries' import substitution policies had similar effects. With freer trade and more intense competition, both the possibility of, and the need for, sourcing investments in manufacturing arose: International corporations started to specialize and rationalize their plants to exploit national comparative advantages. Even where economic and technical conditions prohibited such specialization—for example, for cement, glass, or industrial gases—competitive actions became coordinated across subsidiaries as the companies realized they were competing in a very concentrated global oligopoly. As a result, portfolio foreign investments, where only intangible assets are leveraged, gave way to strategically coordinated integrated operations worldwide, exploiting comparative advantages of different countries for various types of activities. Labor-intensive activities were sited in locations where labor costs were low and from which the world markets were served. Such advantages were most often exploited by owned subsidiaries—through "internalization"—rather than through subcontracting or licensing.[2] This, in turn, led to the development of intrafirm international trade. Such trade may be intraindustry (e.g., the processing of semiconductors overseas for reimport into the United States) or intrafirm but interindustry (e.g., General Electric "offsetting" the sale of jet engines to the Canadian Armed Forces with exports of consumer goods from Canada).

With some significant exceptions—usually government imposed—the trend toward industry globalization and toward MNC integration has affected most countries and most internationally traded goods. The proportion of internationally traded goods in the GNP of countries also increased substantially, so that by 1980 internationally traded goods with substantial trade levels comprised more than 80 percent of the industrial sectors in Western Europe (Orléan, 1986). This trend was particularly strong between 1968 and 1978.

Since the late 1970s, however, three sets of factors have come to limit such globalization. First, the technology no longer always drives toward globalization: New manufacturing techniques may reverse the trend toward "world-scale" plants and allow differentiation and segmentation with smaller cost penalties. Second, protectionism is on the rise and limits the strategic freedom of global competitors. Protectionism applies not only to trade in goods, but also increasingly to trade in knowledge, technology in particular. Third, the organizational and strategic capabilities of global competitors often lag the competitive opportunities available to them, and many firms are less than fully successful in exploiting their opportunities.

The impact of the three sets of limiting factors mentioned above deserves more attention. This chapter reflects this interest, beginning with a selective review of the abundant, if still fragmentary, evidence on the trends toward market homogenization, industry globalization and firm integration, and the underlying forces that drive them. These issues are discussed at three complementary levels of aggregation: the international economic relations framework; individual industries and their competitive dynamics; and the logistics, organizational structures, and management processes of individual firms. Finally, the recent evolution of the three sets of moderating factors—technologies, government policies leading to growing protectionism, and the limited organizational capabilities of firms—and what their effect may be on the fragmentation or globalization of international industries are analyzed.

GLOBALIZATION OF INDUSTRIES

Enabling Conditions

Globalization is rooted in several key enabling conditions: the homogenization of markets, the decreasing costs of transport and communication, decreasing trade barriers, and the competitive pressures from new competitors. First, national markets have become increasingly similar in taste as income distributions in industrialized nations have equalized. The result has been the development of relatively homogeneous market segments that cross borders (Levitt, 1983). Though national markets may have been more similar in the past than was generally recognized (Helleiner, 1981), the media (mainly television), international travel, and the action of active multinational marketers have contributed to the homogenization of markets across national boundaries. Furthermore, global market segments appear in industries as different as automobiles (to the advantage of BMW or AMC's "Jeep") and beer (to the advantage of Heineken and a few others). Higher disposable incomes also encouraged the development of a market for fashionable "world products" in a number of countries, be these products such as British raincoats, Italian sweaters, Swiss watches (Rolex or Swatch), French wines, or Japanese consumer electronics.

Lower communication and transportation costs—the second enabling condition—also made serving these homogeneous markets from centralized locations economical, even for relatively bulky products such as cars. Real-time low-cost communication also made the coordination of a complex worldwide logistics system feasible. The globalization of manufacturing in certain industries where products are complex and differentiated might not have happened without the drastic reductions in transportation and telecommunication costs between 1950 and 1980.

That trade barriers were removed between the 1950s and the 1980s is well known and needs no detailed analysis here. The removal of these barriers provided a third enabling condition for the globalization of industries. Only in some industries where government-controlled customers predominate, and where national defense considerations are relevant, did trade barriers stay in place (Doz, 1986). Specific trade agreements (e.g., the Lomé convention), as well as the extension of credit to developing countries, allowed these countries to participate in this move toward free trade, initiated by traditional industrialized countries. The recognition that across-the-board import substitution measures usually fail, and the successful example of the newly industrialized countries (NICs), also provided an incentive for developing countries to participate actively in the world economy.

A fourth enabling condition, usually at the level of individual firms, was the existence of the organizational infrastructure for globalization. In the mid-1960s, when trade liberalization was initiated and national markets started to converge visibly, many MNCs were already in place, with their infrastructure of sales subsidiaries and foreign plants. This gave them the capability both to gather data worldwide and to respond quickly—at least in theory—to globalization trends. Global information networks and means of global market reach were already in place, decreasing the cost of transition from national to global competition for the major competitors. Experience in handling foreign manufacture, new product introduction, and technology transfer facilitated a prompt response to industry globalization by MNCs (Vernon, 1979). Where such networks and means did not exist, helping hands could be found. Initially, for example, Japanese exporters relied on Japanese trading companies, importing countries' mass merchandisers (e.g., Sears in the United States), mass buyers (e.g., TV rental companies in the United Kingdom), and original equipment manufacturer (OEM) customers (e.g., for computer peripherals). This allowed the new competitors to skip both the market intelligence tasks (Sears, for instance, specified the TV sets it wanted) and the initial market access cost and delay. More complex, more fragmented, less transparent, and less willing distribution structures would have been a formidable barrier to globalization and, where present, remain a source of asymmetry in globalization (witness the painful efforts of many European and American firms to establish a significant market presence in Japan).

Driving Forces for Sourcing and Marketing Globalization

The enabling conditions summarized above were necessary, though not sufficient, for industry globalization to take place. They had to be exploited by firms trying to gain a permanent competitive advantage. The intense competition created by these firms was in most cases the main driving force for integration and globalization to actually take place. Intense competition itself depended on the opportunity for substantial gains through globalization, the existence or the creation of destabilizing conditions, and the presence of competitors with the strategic intent and capabilities to exploit destabilizing conditions to their advantage.

Growing economies of scale in R&D and in production provided the most frequent opportunity for increased profits through globalization. Changes in product and process technology have increased the minimum efficient size of production in a variety of industries, such as cars, chemicals, consumer electronics, semiconductors, and machinery. Combined with the emergence of smaller differentiated global segments, this is a powerful incentive to pool demand from a variety of national markets and serve such demand from large, optimally sited specialized plants. New product development costs have also risen considerably in a range of industries, the best-known of which are aircraft, telephone switching, cars, and semiconductors. These higher costs have created a strong incentive for industries to serve the world market to spread R&D costs over a larger production. There is a further incentive to serve the world market quickly to minimize the financing cost of the initial investment and the competitive risk of technological obsolescence (Hamel and Prahalad, 1985).

In some capital goods industries, such as papermaking machinery, electrical equipment, and railroad equipment, the cyclicality of domestic demand and the uncertainty of future domestic orders have led to chronic overcapacity and to the need for national firms to diversify their customer base by selling abroad. Intense competition, though, is the key driving force. In Europe, following the European Economic Community's (EEC) lowering of trade barriers, little change toward a more efficient industry structure took place unless triggered and stimulated by intense competition (Owen, 1983).

The emergence of a period of intense competition was facilitated by technological or market discontinuities that destabilized the existing market and industry structures. Increases in energy costs, for example, destabilized the structure of such industries as automobiles and papermaking machinery, making it possible for new global competitors to emerge. Shifts from electromechanical to electronic technologies in industries ranging from watches to digital switching systems and avionics have similarly allowed new competitors to establish themselves and occasionally to render a whole industry obsolete (e.g., the mechanical Swiss watch industry). Wide fluctuations in exchange rates have occasionally had significant effects, helping new competitors to penetrate mature industries, even in the absence of new technology or changing economies of scale. In 1983 and 1984, for example, the overvaluation of the U.S. dollar helped Komatsu gain overseas customers at the expense of Caterpillar (Bartlett, 1985).

Ambitious competitors, with a vision of how to turn situations to their advantage, were also needed to make competition more intense. These competitors, such as Komatsu, have the long-term strategic intent to dominate their industry, and they are able to exploit opportunities as they arise. While the new competitors have usually been Japanese, they are also occasionally European (e.g., Leroy Somer in small electric motors) or American (e.g., Otis in the small-elevator industry). Confronted with intense competition from new competitors intent on exploiting economies of scale, new product and process technology, and other destabilizing factors such as exchange rate fluctuations, established competitors have typically reacted in two complementary ways: (1) reducing costs through the exploitation of economies of scale or through economies of location; and (2) gaining worldwide market access through their own efforts or through networks of partnerships and coalitions.[3] First, many companies attempted to reduce costs by exploiting potential economies of large scale, typically by integrating and rationalizing production in a region (e.g., Philips rationalizing its television tube and receiver plants in Europe and Ford doing the same with cars). Companies also searched for lower factor costs and other locational advantages (e.g., the moves offshore in the U.S. electronics industry in the late 1960s and 1970s, or the relocation of the aluminum smelting industry). Demands for cost competitiveness led to sourcing globalization. This practice was sometimes encouraged by governments, either through factor subsidization (small open countries such as Ireland or Singapore) or by the imposition of export performance requirements in exchange for access to local markets, typically in large "promising" countries such as Spain, Brazil, and recently India (Guysinger et al., 1984).

Managing costs is not enough, however, as companies have also come to recognize the value of worldwide market access (Hamel and Prahalad, 1985). Such access is becoming critical not only as a response to rising R&D costs but also as a way of providing potential for competitive retaliation. Goodyear responded to Michelin's inroads into the U.S. tire market by competing more actively in Europe, where Michelin was dominant, thus depriving Michelin of the cash flow it needed to continue its investments to gain market share in the United States. A fight confined to the U.S. market would have been more costly for Goodyear than for Michelin. Similarly, IBM fights Japanese computer manufacturers not so much in the United States, where it would hurt itself, as in Japan, where IBM hurts its Japanese competitors most at the least cost to itself (New Scientist, 1985). U.S. makers of consumer electronics had no such option and fell almost defenseless to the Japanese and to Philips.[4]

Where firms were not yet global enough and could not establish market presence quickly (either because of government restrictions, or because distribution channels are hard to penetrate, or both), strategic partnerships and coalitions developed in industries that were becoming global. The primary motive of most partnerships and coalitions is to shore up market presence and technological competence to establish quickly a defensible position in a global industry. While these do provide a viable option, the sharing of strategic control over competitive actions by several partners usually results in tensions as soon as the external technological and market conditions evolve or the relative strategic importance of the joint activities to the various partners changes. This is probably the single largest cause of mortality in collaborative agreements. Even when the collaboration endures, conflicting priorities may result in delays that blunt its competitiveness (e.g., the 2-year delay in the launch of the A-320 airplane by Airbus Industries and the continuing tensions between the main partners on future product policy and on acceptable financial performance).

Empirical Evidence

Although anecdotal evidence of industry globalization and MNC integration abounds, systematic measurable data on their extent remain scarce and fragmentary. Some industries are well documented (e.g., automobiles, textiles, electronic components, aerospace) through numerous industry studies, but most others are much less well analyzed.[5] Aggregate statistics using proxies such as intrafirm trade also suggest that integration of operations within MNCs is important, with 20 to 30 percent of the international trade of countries such as the United States, the United Kingdom, and Sweden being intrafirm trade. Intrafirm trade seems to be more prevalent in R&D-intensive industries, with high wages and large plants, which is consistent with the driving forces hypothesized above (Dunning and Pearce, 1985; Lall, 1978; United Nations Center on Transnational Corporations, 1983). Yet, even the most detailed studies are fraught with problems in the availability and interpretation of data (Hood and Young, 1980). There is a convergence between findings from studies that start with trade statistics (e.g., based on the U.S. Department of Commerce Annual Survey of U.S. corporations), those that start with a survey of large samples of firms (e.g., Dunning and Pearce, 1985), and those that start with an analysis of the strategic behavior of firms (e.g., Hood and Young, 1980). The more anecdotal evidence from individual "case" studies and from industry-specific studies also points in the same direction.

Industry studies also provide evidence that even in industries that are traditionally nationally fragmented, pressures for integration and globalization are being felt. In the furniture industry, for instance, companies such as IKEA or Habitat-Mothercare are exploiting economies of scale in purchasing, subcontracting, advertising, and brand image, shifting the bottom end of the furniture market away from a fragmented national structure to an integrated multinational one. Similar moves are made at the top end of this market with international designers' brands and with global distributors, such as Roche-Bobois. Even where national prestige, national defense, and strategic independence have traditionally weighed more heavily than competitiveness in industrial choices, original patterns of globalization and integration develop. By and large, European integration in aerospace is making progress under tight supervision from governments. The failure to agree on a single design for a future fighter plane may ultimately be beneficial in offering two complementary products and maintaining spirited competition for export orders: Britain, Italy, Germany, and Spain joined forces and will compete against France and smaller countries. Similarly, the European microelectronics industry is evolving out of a stalemate. We see cooperation between large firms that traditionally were competitors as in the joint development of "megachips" by Philips and Siemens, and we see new ventures occasionally being funded by old firms, as when European Silicon Structures is financed by a group of large European electronic industry firms to make semi-custom chips economically in Europe. Although these collaborative ventures may not operate under the best possible conditions, and their cost of coordination is high, they at least overcome the worst aspects of fragmentation.[6]

Besides the turnaround in many governments in favor of government-sponsored transnational cooperation, it is also important to note that pressure groups that might have tried to block globalization and integration by and large have failed. Although in the early 1970s it seemed plausible that unions would gain a strong say in MNC management, they have now been ruled out as a severe barrier to globalization and integration. This is the result of a combination of factors, namely, the change of attitude in Europe (both the effect of the unemployment crisis and also of an ideological shift away from statism and socialism), the failure of unions to lock MNCs into transnational bargaining, the lack of support provided by governments (e.g., the inability to get the Vredeling proposal off the ground), and the divisive aspects of MNC integration itself on international labor cooperation. Where unions succeeded in gaining a say, as they did with the German codetermination laws, their representatives quickly aligned their positions on those of management.

Economically weak but politically strong national industrial companies could also be barriers to globalization, but by and large they fell to competitive pressures in Europe. Only in a few partly competitive but largely government-controlled sectors, such as electrical equipment for railroads, do the old industry structures survive largely unchanged. Even in some of these industries, there are encouraging signs of possible rationalization, such as the investments by Compagnie Générale d'Electricité into Ateliers de Constructions Electriques de Charleroi. Computer manufacturers are victims of probably the worst stalemate along these lines in Europe. Britain, Germany, and France each have their "national champion," hopelessly small for global competitiveness, and unable to renew its product line without much outside help—usually Japanese. Yet each of these national champions is well enough ensconced in its national political and economic environment to survive, to prevent its merger into a transnational alliance, and to block the development of new, more entrepreneurial national or international competitors. First-class customers desert European suppliers—mainly to IBM—despite the switching costs involved, and the technical capability of European computer companies is withered by their Japanese partners, who provide them with components, critical subsystems, and peripherals. The continuation of this stalemate threatens the European computer industry with extinction. In Europe, though, this is more the exception than the rule, and in most industries—aerospace, chemicals and plastics, pharmaceuticals, and even now automobiles—are taking on the challenge of global competition with a fair measure of success.

Research and Development

Unlike marketing and manufacturing, research and development have not been significantly affected by globalization and have remained principally home-country activities. In a world of sequential market development, where new products and new processes were first developed and put to use on the domestic market or in the home plants, home-country R&D made good sense. As foreign markets developed to resemble the domestic markets, or as foreign plants were built, new products and technologies were transferred abroad once they had been proved domestically. Foreign R&D was mainly devoted to the adaptation of transferred products and processes to local conditions such as taste, product features, norms and standards, and climate (Fischer and Behrman, 1979; Hirschey and

Caves, 1981). The role of foreign subsidiaries was not to innovate on their own but to absorb and apply technology developed in the parent company's laboratories. Technology was in fact global from the start, but research was centrally performed and leveraged internationally through product life cycle phenomena or through transfers to foreign subsidiaries. Even a group such as Brown Boveri, which epitomized the nationally responsive—and fragmented—MNC, leveraged its Swiss-developed technology in its foreign operations (Doz, 1978).

In a more complex world, where the United States no longer clearly leads in product innovation, nor Europe in process innovation, centralized R&D is less effective. First, the leading users—those who can contribute their experience to the success of an innovation—are no longer necessarily available in the domestic market. Although this is truest for MNCs in small countries (e.g., Holland, Sweden, Switzerland, Korea), it is also applicable to U.S. or Japanese companies. Leading markets for medical electronics, for example, may be in the United States and in Japan and Sweden for factory automation, in France for nuclear engineering, and in Britain for consumer electronics. Second, key scientists, like the leading users, are potentially more dispersed geographically than they have ever been. Some European pharmaceuticals or electronics firms find it easier to locate laboratories for new technologies such as genetic engineering or microchips in the United States than in Europe. Conversely, India may offer the potential for a large number of inexpensive software specialists and Italy for creative ones. Several U.S. electronics companies, such as Control Data, Motorola, and Texas Instruments, are setting up software R&D centers in India and Italy. Exploiting a larger pool of talent and avoiding the cost of expatriation are strong motives to locate R&D in various countries. Third, locating R&D in host countries may also help placate their governments' desires for more higher skilled jobs (see Branscomb in this volume).[7] It may also make the firm eligible for national R&D subsidies or access to national collaborative projects. Finally, the mobility and transfer of knowledge within MNCs is neither easy nor costless (Teece, 1977).

Despite these trends, the forces favoring centralization of R&D remain strong. First, as markets become increasingly homogeneous, the need for specific local product adaptation or for autonomous product development is lessened. Second, the benefits of close proximity of researchers are strong. Although estimates of the distance beyond which easy informal communication between scientists breaks down range from a few yards to a 1-day plane commute, observers agree that the scattering of related research activities is detrimental to their effectiveness (Allen, 1977). Third, there are often economies of scope in R&D, particularly where technologies are interdependent, which make the scattering of R&D laboratories costly since they cannot be made self-contained. Fourth, considerations of political risk seem to have limited the willingness of major firms to be dependent in their home markets on technologies developed abroad.

It is thus no great surprise to observe that, with a few notable exceptions, the performance of the R&D function in firms has neither been globalized nor integrated to any extent comparable to that of manufacturing and marketing. The results of R&D are global, not the performance of the R&D tasks. With a few significant exceptions, R&D remained centralized, at least in the technology-intensive sectors. Foreign R&D labs do mostly product development and adaptation to local conditions, or sometimes basic research, but seldom have broad research mandates, except in some of the most mature MNCs (e.g., IBM, Dow Chemical Company, and Ciba Geigy). Further, few firms seem to have developed systematic processes for the coordination of R&D activities across regions of the world, again with a few notable exceptions, such as IBM.

Yet, as the home market can no longer be equated with the lead market, centrally performed R&D needs to be responsive to the needs of distant potential users. This may be easy to achieve for engineered commodities, such as consumer durables, photocopiers, and typewriters, but it is more difficult where needs can be defined only in close conjunction with users rather than through market research (von Hippel, 1982). Whereas Japanese successes have been confined mainly to engineered commodities, European exporters and MNCs cover a wider spectrum of products. Large European companies have particularly difficult problems with their U.S. subsidiaries. Their products are often developed with too much of a "technology-push" by central labs whose scientists and managers may have gained a sense for European needs but are insensitive to U.S. needs. In some cases, they seem to be following what they think is the "right" path from a technological rather than market standpoint without considering the lead users' needs. As a result, it is not uncommon for U.S. subsidiaries of European groups to avoid marketing products developed in Europe, thus ensuring that their volumes will be too low to break even. Instead, they develop new products at great cost, take a license from a competitor, or buy the products directly on an OEM basis.

The issue is not who is right or wrong between the U.S. subsidiary and headquarters, but the fact that a European group facing such a situation gains little competitive advantage from being in the United States at all. The converse example, of insensitivity by U.S.-based companies to non-U.S. market needs in their product development, is better known and more easily explicable, given the historical dominance—in the operation of most U.S. MNCs—of the U.S. market over smaller fragmented national markets. Faced with the dilemma between economies of centralization and the market access advantages of R&D dispersion, MNCs have occasionally done both; some U.S. MNCs have maintained central laboratories but located the primary labs for a set of products in the lead market away from headquarters.

In summary, R&D activities have not changed as dramatically over time as manufacturing and marketing: Their activities have remained largely centralized—most often in the home country—and their output leveraged through transfer to foreign subsidiaries or through embodiment in exported products (Hirschey and Caves, 1981). For most MNCs, the arguments for centralization seem to have outweighed those favoring geographical fragmentation.

Summary Observations

Although it has to remain impressionistic, since detailed data are lacking, the analysis of the balance between forces of global homogenization and integration and forces of fragmentation clearly shows the balance tilting toward globalization. The removal of trade barriers, and the growing similarity of national markets created the potential for globalization of markets and competition. The development of MNCs, or of global networks allying independent firms, and the technology of cheap effective transportation and communication provided the practical means necessary for the integration of supply. These conditions were necessary, though not sufficient. Intense competition in most industries was the driving force necessary for integration and globalization. During the same period, actors who might have stalled globalization either did not act or acted ineffectively.

Thus, homogenization of markets has increased, industries have globalized, and firms have responded by geographic integration of their activities. Such integration took place (1) for sourcing—usually driven by manufacturing cost-reduction opportunities stemming from growing economies of scale and from economies of location; (2) for marketing—usually driven by a mix of economies of scale in distribution and manufacturing; and (3) by the competitive leverage brought by market scope. Coalitions and partnerships of all kinds provide an attractive low-cost alternative to single-firm manufacturing and market access investments. They do not, however, provide the strategic freedom and control available through a company's own investments.

Research and development activities have typically remained in the home country. However, as more and more products are developed for world markets, usually for simultaneous rather than sequential introduction, the need for a better integration of foreign subsidiaries and domestic labs has arisen. Evidence from specific product innovation studies in the United States and in Europe tends to suggest that this need for integration is not well met. Conversely, there is little to suggest that MNCs successfully apply innovations that originate in one subsidiary outside of that subsidiary.

The next section discusses three sets of factors that suggest the trends toward homogenization, globalization, and integration may slow down or even reverse themselves in the coming decade. Some of the underlying conditions or driving forces will have run their course, and new limits may appear.

LIMITS TO GLOBALIZATION

Manufacturing Technology

The evolution of manufacturing technology—in particular the increase in economies of scale in manufacture—has been one of the key conditions in favor of market globalization and MNC integration in a number of industries. Several factors may now slow down this trend. First, new technology has been so successful at reducing manufacturing unit costs that these costs now account for only a small proportion of total delivered costs. Further reduction of manufacturing cost will be of lesser impact than in the past, as other elements of cost play a much greater role, namely overheads, R&D recovery, and distribution.

Second, economies of scale may no longer increase in the same way as in the past. Some new technologies may abruptly decrease economies of scale. New multipurpose smaller processors in the chemical industry are an example of this type of technology. Even in the absence of genuinely new technology that would reduce economies of scale, the advantage of manufacturing systems—from the well-known materials and resource planning systems to the embryonic "factory-of-the-future" concepts—are based on cost reduction from better managing the manufacturing system rather than from increasing the plant size or the length of the production run. Better manufacturing processes allow more flexibility in production. For instance, multiple car models can be produced in varying proportions on the same assembly line with relatively little cost penalty. This could allow car manufacturers to move back from large single-model factories serving multicountry markets to multimodel factories serving single-country markets. Although there may still be some cost penalty to setting up a flexible factory rather than a narrowly focused one, at least the trade-off between increasing flexibility and decreasing costs can be explicitly considered.

The impact of flexible manufacturing on the trade-off between integration and fragmentation of manufacturing is still unclear, however. Greater flexibility allows producers to cater to shifts in consumer preference—as their discretionary income increases—from cheap standardized goods toward customized products. Flexible manufacturing systems may allow both product customization—at least so long as such customization can be achieved through featurization around a common core—and low cost. These systems may shift the basis for cost advantage from scale to scope and thus make it possible for an integrated manufacturer to serve differentiated worldwide needs.

Third, the "just-in-time" manufacturing concept works best with the colocation of various facilities into an integrated system. This polarizes globalization and integration to the extremes: either a series of small "local-for-local" plants, each by and large self-sufficient, or, at the opposite extreme, a single integrated source for everything (e.g., Toyota City, or to a lesser extent, Boeing around Seattle, or Caterpillar around Peoria). A widely dispersed integrated manufacturing network (such as Ford of Europe), where plants are distant and supply each other with components and subassemblies, is least amenable to just-in-time manufacturing management. Buffer inventories must be kept to allow for transportation delays, localized strikes and disruptions, and slowdowns in custom clearance. This would suggest that the initial patterns of integration within MNCs, particularly in Europe, may not endure. Either the advantages of colocation and flexibility will be such that we will witness a return to largely local plants, or the advantages of focus and specialization will continue to exceed those of flexibility, and the advantages of collocation will lead to even further centralization of manufacturing.

Fourth, the trends toward vertical deintegration may allow more creative combinations between independent firms at different stages in a value-added structure. This would allow producers to continue to draw benefits from economies of scale for components and to gain flexibility for end products. Large-scale component manufacturing can be delegated to independent suppliers serving multiple smaller-scale assemblers, for instance. This may lead to different balances between integration and fragmentation at various stages of the value-added chain in the same industry.

Most industries and firms are not yet affected by all of these trends, but economies of scale in production are unlikely to be the opportunity they were in the 1960s and 1970s. As a result, economies of scale will no longer be a driving force toward globalization and integration. Choices for MNC managers will be more complex than just building up the largest plants with the aim of regaining competitiveness. Most companies are likely to end up with a mix of plants of various sizes and locations, and with various degrees of focus or flexibility.

Economies of location are also likely to become less important. With a few exceptions—such as aluminum—economies of location derive mainly from labor cost advantages. Several observations can be made. Not only has manufacturing cost decreased in relation to delivered cost for a whole range of industrial products, but also labor costs will decrease in relation to manufacturing costs with any shift toward more capital-intensive technologies. Stable or increasing real wages in Europe, despite the recent recessions, have accelerated the substitution of capital for labor. Even with relatively low wages, the product quality provided by automation in consumer electronics, for instance, may lead to rapidly decreasing labor content and to the repatriation and automation of plants previously dispersed from developed countries.

Locations with low labor costs also tend to catch up with locations with higher costs if only because skilled labor is scarce and the general wage structure moves up. Location advantages based on cheap labor are thus often temporary. Although labor may remain cheap in countries where political risks, government policies, or financial problems deter foreign investors—and thus limit the competition for labor—countries such as Singapore, Korea, and Taiwan, which have been hosts to massive foreign investments, have often seen their real-term wage rates increase significantly. In some industries—such as garment production—firms may shift their manufacturing locations in a search for cheaper labor. Where developed countries' firms subcontract to local producers—a prevalent practice for garments—shopping around for cheaper subcontractors is easy; when the foreign MNC sets up its own sourcing plants, however, closing down and relocating elsewhere is a much more costly and difficult process.

Differences in the cost of capital between countries also tend to decrease as the world's capital market becomes more integrated and as MNCs cross-finance themselves on multiple markets and arbitrage between them. Although domestic firms may still benefit from favorable institutional arrangements, e.g., the institutional structure of Japanese capitalism, or from specific government assistance, e.g., European exporters, these advantages are limited, not always accessible to MNC subsidiaries, and not often sufficient to justify location.[8]

Finally, exploiting economies of location also entails certain risks, for instance, exchange risks. If the mix of manufacturing locations differs significantly from that of selling locations, the firm is exposed to currency risks. Whereas this can play in their favor occasionally (e.g., the hefty margins made by European companies exporting·to the United States in 1984-1985), it can also play the other way around as in the plight of U.S.

exporters. Various hedging approaches can be adopted, but they usually either run counter to the search for economies of location, or they result in the creation of abundant "buffer" excess capacity. Instability of the exchange rate only increases the difficulties and costs of these approaches.

Protectionism

Since 1975 protectionist pressures on the U.S. Congress have increased largely as a result of the globalization process. Outright protectionist bills have been avoided only by successive administrations' careful negotiation of selected "voluntary" protection. Examples include the "Orderly Marketing Agreements" for TV sets and the "trigger prices" for steel or other commodities. Proposals such as the Burke-Hartke Act, which would have considerably limited the opportunity for U.S. firms to import goods made by their overseas subsidiaries, have been turned down, but at an increasing political price. The overvaluation of the dollar in 1983-1985, and the huge U.S. trade deficit only made matters worse. In the fall of 1985, only the shift in the U.S. position toward an active intervention policy to devalue the dollar staved off strong protectionist measures.

Europe, while making only slow progress toward a true free internal market, has resorted to protectionism toward a variety of industries, particularly those threatened by Japanese imports. Government purchasing policies that favor national suppliers also endure and close whole industries to foreign suppliers. Whether the Japanese market is closed or just hard to enter is an old debate, but it is clear that market access to Japan in critical industries is extremely difficult.

What is important here is not so much the exact extent of protectionism, but that recent evolutions do not allow managers to make a safe assumption about freer trade. The risk of a widespread return to protectionism puts a damper on globalization strategies that imply high levels of trade and adds fuel to strategies that return to traditional foreign investment as a way of overcoming trade barriers. Indeed, the purpose of many of the Japanese investments in Western Europe and in the United States is to overcome trade barriers, or at least to serve as "insurance" against new trade barriers, should they be implemented.

Among the less-obvious aspects of protectionism that may hamper MNC integration strategies are the issues of data flow across borders. Several countries have argued that data should be likened to raw material and processed locally rather than internationally. The issues are manifold and vary from country to country. Among the most prominent are (1) the importance of local data processing for stimulating the national demand for electronic data processing hardware and services and for telecommunication services; (2) the disadvantage of local firms and governments in relation to MNCs and their access to global market information; and (3) the threat of more centralization of decision making in MNCs, a process directly related to integration strategies. Canada has clearly articulated concerns about transborder data flows. Brazil and France have followed suit, with somewhat different concerns and priorities. Although policies on data flow are often lent moral legitimacy by being amalgamated with a series of regulations to protect the privacy of individuals, economic and political considerations drive the development of such policies. Countries do compete for the location of data processing centers by MNCs, and they also compete for international data transmission and value-added services. A few countries, including the United States and Britain, have taken an aggressive commercial position by lowering packet-switching charges, for example, and others try to regulate data flows. Although the current impact of data flow regulations is limited, it is a concern for at least some firms.[9] In addition, regulation of data flows may also be a way to ensure that critical knowledge exists within the country. One widespread concern, for instance, is that some U.S. suppliers of computers keep debugging software at home, where it can only be accessed by telephone lines from Europe. Should denial measures be taken by the U.S. government, whatever the reasons (as was done in 1982 in the Dresser case), such critical software might no longer be available.[10]

More broadly, protectionism in technology has become a major issue. In the 1980s the U.S. government, as well as several U.S. firms, became worried about the transfer of technology to Japan and to the USSR. This concern arose as the extent and success of efforts by Japanese firms and the Soviet government to appropriate Western technologies became clear. With regard to Japan, the issue is competition, particularly in industries such as semiconductors. In this industry, in particular, manufacturing equipment is critical to success, and the U.S. industry became concerned that process technology was transferred too easily to Japan. The concern was heightened as Japan came to be seen in the United States as an "unfair" competitor. Similar concerns have been voiced in other industries, such as aerospace and computers, as evidenced by IBM's actions against Mitsubishi and Hitachi.

With regard to the USSR, the issue is twofold: first, to deny the USSR access to the core technologies of military systems, a priority widely shared in the West; and second, to limit the USSR's access to technologies that may allow faster economic growth and thus make large military expenditures more affordable to the Soviets. The second point is a matter of debate between U.S. government hard-liners and more liberal circles in the United States and among European governments. The issue gained prominence with the discovery, probably by French counterespionage in early 1982, of the magnitude of the Soviet effort to spy on the West, and of the success of that effort.[11] Later updates, based on captured Soviet documents, kept the issue salient. Also giving prominence to the issue were several instances of discreet reexport of classified U.S. equipment via Sweden and Austria and several cases of industrial espionage in major West European and American companies, including MBB, Dassault, and Hughes. Although studies suggest that the Soviets are not able to absorb and finance the use of the new technology they obtain from the West, legally or otherwise, the Reagan administration took it to heart to stem the flow of technology to the Soviet bloc (Bornstein, 1985).

The Export Administration Amendment Act of July 1985 extends the list of goods subject to U.S. export licenses to "dual-use" equipment, civilian in principle, but using technologies or components with potential military use. The U.S. policy of reexport control also considerably limits the mobility of components to be incorporated into systems assembled in another country, and sold in yet another. European integrated MNCs, such as Philips, suffer great logistic complications from this new set of laws (Dekker, 1985). This leads them to substitute, where possible, non-U.S. for U.S. components and subsystems. Although such substitution is a boost to some European industries, it leads to an inefficient duplication of effort between the United States and Europe.

Protectionism in technology—be it through limiting the transfer of data or through restrictions on exports of goods possibly related to the manufacture of defense-related equipment—makes it difficult for technology-intensive MNCs to adopt integration strategies, since the various parts of the company need to be technologically autonomous. It also makes it difficult for U.S. firms to cooperate with foreign partners on joint R&D and casts doubt on the ability of European firms to use technology they would have acquired through collaborative efforts with the United States or with U.S. government support. Although Japanese firms are more strongly encouraged than their European counterparts to participate in U.S. defense projects, the same issues arise between the United States and Japan as between the United States and Europe. Conversely, IBM's or Texas Instruments' access to the results of joint Japanese research projects is a difficult issue.

These concerns have prompted Europe into action, first with the European Strategic Program for Research and Development in Information Technology (ESPRIT) and with specialized projects, such as Research in Advanced Communications in Europe (RACE) and more recently with a program called EUREKA. ESPRIT's relative success was a surprise, but by the end of 1985 about 195 projects shared 1.4 billion European Currency

Units, and many of them looked promising. EUREKA, launched as a civilian equivalent to the U.S. Strategic Defense Initiative, is still embryonic and funding is uncertain. Despite widespread skepticism, it may take hold and lead to interesting projects. This direct subsidy approach addresses only one facet of European competitiveness, however, and maybe not the most important: European firms show inferiority not in the development of new technology but in its exploitation. Technology may not be the critical issue. Market structure and management are. Although much attention in Europe is focused on making Europe a true "common" market, remarkably little attention is devoted to managerial limits to the successful exploitation of global technological and competitive opportunities.

Organizational Capabilities

The various elements discussed above suggest that large international competitors will face a world of neither fragmentation nor global integration, but a mixture of both, with many shades of gray and complex patterns of international operations that are unlikely to fall neatly into any category. Thus, there will be many trade-offs between industry fragmentation and globalization and strategies of integration and subsidiary autonomy, and they will vary by function, country, and business. Differences between industries, between segments within the same industry, and even between stages in the value-added chain are going to be important. This will introduce considerable variety in the situations faced by MNCs. Further, strategies will vary from free and competitive to negotiated and collaborative through complex networks of collaborative agreements, coalitions and joint ventures among firms, and occasionally between them and governments (Doz, 1986).

Not all global competitors are able, organizationally, to cope with such diversity. Most started as national companies (e.g., most Japanese competitors) or with fragmented organizational structures loosely "federated" by headquarters. Such fragmented structures, leaving a lot of autonomy to individual subsidiaries in various countries, fit well with the fragmented environment faced by MNCs prior to the 1970s.

The initial transition from autonomous subsidiaries to coordinated international strategies and integrated manufacturing and marketing networks has been a traumatic experience for many companies. The process has been slow (typically 3 to 7 years), painful, and not always successful (Doz and Prahalad, 1981; Prahalad and Doz, 1981). For a while in the mid-1970s, matrix organizations were seen as the answer to complex trade-offs between integration and fragmentation. Though a matrix organization may achieve such trade-offs, it achieves them well only if a number of conditions are met.

First, a matrix organization is not merely a different form of organization. Rather, it is a different mode of making decisions and ensuring that relevant data and perspectives are brought to bear on the choices, that trade-offs are made explicit, and that well-considered decisions are reached. This requires both a well-developed management system infrastructure, the involvement of top management, and much attention to the quality of the executive process. Observations of many companies suggest that not all are able to meet these conditions. Hence the widespread disillusionment with matrix organizations (Prahalad and Doz, 1987).

Although the "ideal" MNC organization is easy to spell out in principle, it is difficult to put in place and make work. Yet, as discussed in the earlier sections, the conflicting demands for flexibility and responsiveness, on the one hand, and for global competitiveness and integration, on the other, call for complex trade-offs. Such conflicting demands thus further limit the capabilities of firms to succeed in global industries.

Moreover, in many industries, speed and interdependence in action become increasingly critical. Product cycles are shorter, and the maintenance of competitive advantage requires coordinated policies across product lines and business units, both for technology development and for market access (Hamel and Prahalad, 1985). The growing number and variety of collaborative arrangements also make it more difficult for companies to maintain conventional configurations of strategic control, as can be more easily done with fully owned operations (Doz, 1986). As a result, a gap develops between the demands put on companies by global competition and the capability of their organizations and management to meet them.

CONCLUSION

The three sets of factors outlined above—manufacturing technology, protectionism, and organizational capabilities—may limit the growth of integrated multinational companies and tilt the balance again toward fragmentation. Collaborative agreements and strategic partnerships may increasingly represent an alternative to direct investment for gaining market access, achieving volume production, or leveraging technology. These may deeply modify the nature of global competition and international industries by creating a series of intermediate positions between national and global competitors.

NOTES

1. For summary data, see Dunning and Pearce, 1985; Stopford, 1983; Vernon, 1977; Franko, 1976. See also, for U.S. multinationals, U.S. Bureau of Economic Analysis, 1986.
2. For a summary of the internalization argument, see Casson, 1979; Rugman, 1981; Dunning, 1979. Many authors draw on Hymer, 1976.
3. For a general argument on the dynamics of global competition, illustrated with the example of color television sets, see Hamel and Prahalad, 1985.
4. See Hamal and Prahalad, 1985, for a summary argument. For a more detailed analysis, see Millstein, 1983.
5. For a series of industrial studies, see Zysman and Tyson, 1983; Hochmuth and Davidson, 1985.
6. For an early analysis of these problems in collaborative ventures, see Hochmuth, 1974.
7. The argument cuts both ways, though, as it may be argued that local scientists or technicians employed by MNCs develop knowledge, the economic benefits from which may well accrue to another country where the MNC operates, whereas local firms would have a greater propensity to export innovative goods and processes, thus creating more value for the country.
8. For a more detailed discussion of the limits to the competitive advantage that can be obtained from multinational resource deployment, see Doz and Prahalad, 1986.
9. For a summary analysis, see Kane, 1985, and United Nations Center on Transnational Corporations, 1982.
10. For a detailed discussion of the Dresser case, see Bettis, 1984.
11. Although not publicly available, the various CIA reports to the U.S. Congress did much to increase the political salience of the transfer of technology to the Soviet Union.

REFERENCES

Allen, T. A. 1977. The Flow of Technology. Cambridge, Mass.: MIT Press.

Bartlett, C. 1985. Komatsu Limited. Harvard Business School Case Study. HBSCS 0-385-277.

Bettis, R. A. 1984. Dresser Industries and the Pipeline. Southern Methodist University Case Study.

Bornstein, M. 1985. East-West Technology Transfer: The Transfer of Western Technology to the USSR. Paris: Organization for Economic Cooperation and Development.

Casson, M. 1979. Alternatives to the Multinational Enterprise. London: Macmillan.

Dekker, W. 1985. The technology gap: Western countries growing apart. Speech presented at the Atlantic Institute for International Affairs, Paris, December 5, 1985.

Doz, Y. 1978. Brown Boveri & Cie. Harvard Business School Case Study, HBSCS 4-378-115.

Doz, Y. 1986. Government policies and global competition. In M. E. Porter, ed., Competition in Global Industries. Boston: Havard Business School Press.

Doz, Y., and C. K. Prahalad. 1981. Headquarter influence and strategic control in multinational companies. Sloan Management Review 23(1).

Doz, Y., and C. K. Prahalad. 1986. Quality of management: An emerging source of global competitive advantage? In N. Hood and J. E. Vahlne, eds., Strategies in Global Competition. London: John Wiley & Sons.

Dunning, J. 1979. Explaining changing patterns of international production: In defense of the eclectic theory. Oxford Bulletin of Economics and Statistics 41(November):269-296.

Dunning, J. H., and R. D. Pearce. 1985. The World's Largest Industrial Enterprises, 1962-1983. New York: St. Martin's Press.

Fischer, W. A., and J. N. Behrman. 1979. The coordination of foreign R&D activities by transnational corporations. Journal of International Business Studies 10-3(winter):28-35.

Franko, L. G. 1976. The European Multinationals. Stamford, Conn.: Greylock.

Guysinger, S., et al. 1984. Investment Incentives and Performance Requirements. Washington, D.C.: The World Bank Mimeographed Report.

Hamel, G., and C. K. Prahalad. 1985. Do you really have a global strategy? Harvard Business Review (July-August):139-148.

Helleiner, G. K. 1981. Intra Firm Trade and the Developing Countries. New York: St. Martin's Press.

Hirschey, R. C., and R. E. Caves. 1981. Research and the Transfer of Technology by Multinational Enterprises. Oxford Bulletin of Economics and Statistics 43(2):115-130.

Hochmuth, M. S. 1974. Organizing the Transnational: The Experience with Transnational Enterprise in Advanced Technology. Cambridge, Mass.: Harvard University Press.

Hochmuth, M. S., and W. Davidson. 1985. Revitalizing American Industry. Cambridge, Mass: Ballinger.

Hood, N., and S. Young. 1980. European Development Strategies of U.S.-Owned Manufacturing Companies Located in Scotland. Edinburgh: Her Majesty's Stationery Office.

Hymer, S. 1976. The International Operations of National Firms: A Study of Foreign Investment. Cambridge, Mass: MIT Press.

Kane, M. J. 1985. A Study of the Impact of Transborder Data Flow: Regulation on Large U.S.-Based Corporations Using an Extended Information Systems Interface Model. Ph.D. dissertation. College of Business Administration, University of South Carolina.

Lall, S. 1978. The pattern of intra firm exports by U.S. multinationals. Oxford Bulletin of Economics and Statistics 40(3):209-223.

Levitt, T. 1983. The Globalization of Markets. Havard Business Review (May/June):92-102.

Millstein, J. E. 1983. Decline in an expanding industry: Japanaese competition in color television. In J. Zysman and L. Tyson, eds., American Industry in International Competition. Ithaca, N.Y.: Cornell University Press.

New Scientist. 1985. IBM begins its Japanese assault. (17 October):22-23.

Orléan, A. 1986. "L'insertion dans les échanges internationaux: comparison de cinq grands pays développés. Economie et Statistiques 184(Janvier):25-39.

Owen, N. 1983. Economies of Scale, Competitiveness and Trade Patterns Within the European Community. Oxford: Clarendon Press.

Prahalad, C. K., and Y. Doz. 1981. An approach to strategic control in multinational companies. Sloan Management Review 22(4):5-13.

Prahalad, C. K., and Y. Doz. 1987 (forthcoming). The Multinationals'Mission. New York: The Free Press.

Rugman, A. M. 1981. Inside the Multinationals: The Economies of Internal Markets. London: Croom Helm.

Stopford, J. M. 1983. The World Directory of Multinational Enterprises, 1982-83. London: MacMillan.

Teece, D. J. 1977. Technology transfer by multinational firms: The resource cost of transferring technological know-how. The Economic Journal 87(June):242-261.

United Nations Center on Transnational Corporations. 1982. Transnational Corporations and Transborder Data Flows: A Technical Paper. New York: United Nations.

United Nations Center on Transnational Corporations. 1983. Transnational Corporations in World Development: Third Survey. New York: United Nations.

U.S. Bureau of Economic Analysis. 1986. U.S. Direct Investment Abroad: 1982 Benchmark & Survey Data. Washington, D.C.: U.S. Government Printing Office.

Vernon, R. 1977. Storm Over the Multinationals. Cambridge, Mass.: Harvard University Press.

Vernon, R. 1979. The product cycle hypothesis in a new international environment. Oxford Bulletin of Economics and Statistics 41(4).

Vernon, R., and W. H. Davidson. 1979. Foreign Production of Technology-Intensive Products by U.S.-based Multinational Enterprises. Harvard Business School Working Paper, HBS 79-5.

von Hippel, E. 1982. Appropriability of innovation benefit as a predictor of the functional locus of innovation. Research Policy 11(2):95-115.

Zysman, J., and L. Tyson, eds. 1983. American Industry in International Competition. Ithaca, N.Y.: Cornell University Press.

The Impacts of Technology in the Services Sector

JAMES BRIAN QUINN

In recent years much attention has been appropriately focused on the structural changes technology has wrought upon manufacturing, particularly in the United States. But technology has created even more dramatic changes in the services sector, which now accounts for some 68 percent of U.S. GNP and 71 percent of U.S. employment. The shift toward services has been a long-term trend, not only in the United States but in all major industrialized countries (see Figures 1 and 2). This chapter addresses a variety of issues relating to technological change and services:

- What are the major causes and implications of the shift toward a services economy?
- How has technology restructured the services sector and how does this restructuring affect U.S. trade and competitiveness?
- Can a services economy generate a continuously higher standard of living?
- How might a services economy and services technologies affect national sovereignty and the nation's posture in the world?

WHAT IS THE SERVICES SECTOR?

Many engineers and executives mistakenly perceive the services sector as "making hamburgers" or "shining shoes." Such simplifications belie the complexity, power, technological sophistication, and continuing growth potentials of services in a modern economy. Although there is not complete consensus on a definition for the services sector, it is generally considered

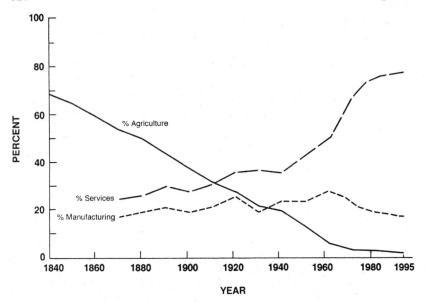

FIGURE 1   Employment in industrial sectors as percentage of total labor force. From Quinn (1983).

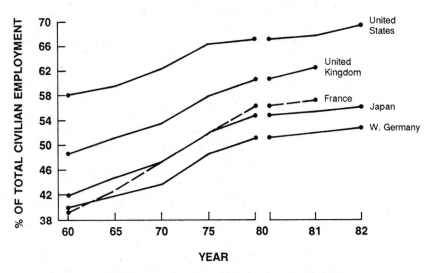

FIGURE 2   Percentage of employment in service industries, five nations, selected years, 1960 to 1982. From Quinn (1986).

to embrace all those Standard Industrial Classification categories in which (1) the primary output is not a product or a construction, (2) value is added principally by other means (such as convenience, amusement, feelings of well-being, improved knowledge, security, health, comfort, location availability, or flexibility) that cannot be inventoried, and (3) outputs are essentially consumed when produced (Collier, 1983; Mark, 1982).

"Services" cannot be viewed as a single sector easily isolated from all others. Table 1 suggests its scale, diversity, and economic impact.

If one defines an industry as a group of enterprises whose outputs are largely substitutable for each other, there is no such thing as a single services sector. The sector is at least as heterogeneous as manufacturing; one should think of it as a grouping of diverse industries, just as manufacturing is. This context makes it easier (1) to understand some of the more important relationships between technology and specific services industries, and (2) to dispel some of the myths about the services sector.

MYTHS ABOUT THE SERVICES SECTOR

The first myth about services is that they are somehow less important on a "human needs scale" (Maslow, 1954, chapters 5 and 8) than products. Hence (so the argument goes), services cannot provide the same value added or economic stability as a production economy. In elemental societies, the first value added is created by the mere presence or adequacy of a product, for example, producing food for sustenance, housing for basic shelter, or clothing for protection from the elements.[1] But as soon as there is even a localized self-sufficiency or surplus in a single product, added value is created by distribution of the product—a services activity. In fact, the added production is worthless without the distribution.

In such primitive societies other "services," such as health care, education, trading, entertainment, religion, creative designs, and nonfunctional art works, quickly become more highly valued than basic products, whose production soon becomes the work of the poor. Similarly, in modern societies most of a product's added value is due to service functions: better design, convenience in use, packaging, distribution, marketing presentation, post-purchase serviceability, and so on. And many of the most highly valued (high-priced) activities in the economy are services such as architecture, art, health care, entertainment, travel, banking, investment, personal security, or education.

Initial analyses indicate that measured value added in the services sector is at least as high as in manufacturing (see Figure 3). Services also appear less cyclical than manufacturing. In the last two decades, services employment has advanced an average of 2.1 percent during economic con-

TABLE 1   Components of United States Gross National Product

Services	Current $ Billions					1972 $ Billions				
	1970	1975	1980	1983	1984	1970	1975	1980	1983	1984
Total GNP	993	1,549	2,632	3,305	3,663	1,086	1,234	1,475	1,535	1,639
Agriculture, forestry, fisheries	29	53	77	73	91	34	37	40	39	45
Manufacturing	252	358	582	685	776	261	290	351	354	391
Transportation	39	55	99	115	130	43	46	52	47	50
Communication	24	40	67	92	103	26	36	52	59	63
Wholesale trade	68	117	190	229	265	72	88	104	114	130
Retail trade	98	149	238	307	337	104	122	142	152	165
Finance, insurance, real estate	142	216	399	543	598	155	188	236	254	265
Other service industries	114	186	342	478	529	127	148	189	207	219

tractions and 4.8 percent during expansions. Employment in the goods-producing sector declined an average 8.3 percent in recessions and increased an average 3.8 percent in expansions (Office of the U.S. Trade Representative, 1983, p. 21) (see Figure 4). This means that people will give up many product purchases (indicating lower marginal utility for those products) before they will sacrifice desired services like education, telephones, banking, health care, police, or fire protection.

A second myth is that service industries are much more labor-intensive and less technologically based than manufacturing. Stephen Roach of Morgan Stanley & Company, Inc., has shown that capital stock per services worker has been rising since the mid-1960s and now surpasses that for manufacturing workers (Roach, 1985). Kutcher and Mark's data—grouping 145 industries on the basis of capital stock per worker and labor hours per unit of output—show wide variations in labor intensity in both the manufacturing and services sectors. Some service industries—notably rail and pipeline transportation, broadcasting, communications, public utilities, air transport—are among the most capital-intensive of all industries. Nearly half of the 30 most capital-intensive industries were services. But,

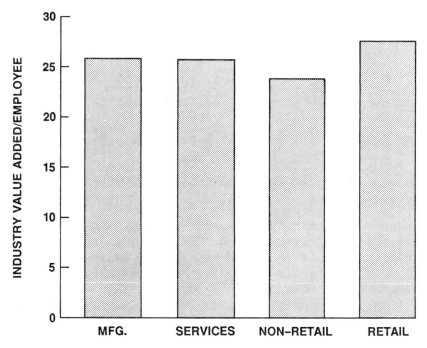

FIGURE 3   PIMS index of value added. Derived from industry survey and calculations from PIMS (Profit Impact of Management Strategy) 1985 data base, Strategic Planning Institute, Cambridge, Massachusetts.

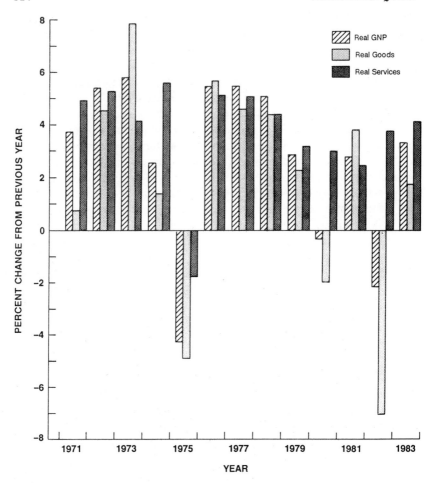

FIGURE 4   Recession resistance of the services sector.

surprisingly, few service industries were found in the three lowest capital-intensity deciles (Kutcher and Mark, 1983). Profit Impact of Management Strategy (PIMS) data from the Strategic Planning Institute, Cambridge, Massachusetts, also show that aggregate capital intensities in services are comparable to those in manufacturing (see Figure 5).

Many service industries are very technology-intensive today. One thinks first of communications, information services, health care, airlines, and public utilities. But the banking, education, financial services, entertainment, car rental, message delivery, and retailing industries have also become technology-intensive (Office of Technology Assessment, 1984). For example, retail discounting (largely by major chain retail operations)

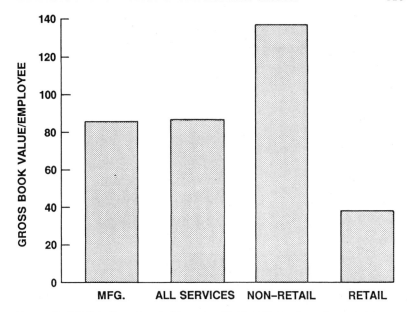

FIGURE 5   PIMS indices of capital intensity. PIMS 1985 data base, Strategic Planning Institute, Cambridge, Massachusetts.

represented 48 percent of all retail general merchandise sold to the public in 1984, and the top five merchandise chains had more than $70 billion in total sales in 1985.[2] These and the large food-retailing chains require extraordinarily sophisticated computer, communications, product control, credit, and cash-management systems to compete successfully. The services sector is a major market for high technology—one study indicating that 80 percent of the computing, communications, and related information technologies equipment sold in 1982 went to the services sector (Kirkland, 1985). In Britain 70 percent of all computer systems sold in 1984 went to the services sector (*The Economist*, July 6, 1985).

A third myth about the services sector is that it is much too small-scale and diffuse either to buy major technological systems or to do research on its own. Again, initial analysis of the PIMS data suggest this is not true. Although detailed Herfindahl indexes are not available, concentration and mechanization in the services sector (see Figures 6 and 7) appear to be about as high as in manufacturing.[3] Thus, the sector has the potential not only to purchase technology but also to contribute to its conception, design, and development. We have not included government agencies or municipalities in our statistics, but they clearly have similar capabilities.

A fourth myth is the fear that a services economy cannot continue to

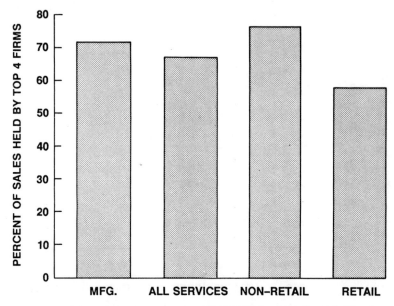

FIGURE 6   PIMS indices of concentration. PIMS 1985 data base, Strategic Planning
Institute, Cambridge, Massachusetts.

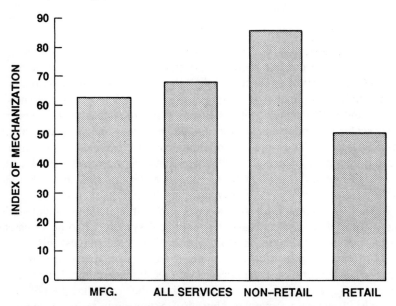

FIGURE 7   PIMS indices of mechanization. Index of mechanization calculated as
[Gross Book Value of Plant and Equipment]/[(Value Added/Net Sales) × (Net Sales
+ Change in Inventory/Percent Plant Utilization)]. From PIMS 1985 data base,
Strategic Planning Institute, Cambridge, Massachusetts.

TABLE 2   Productivity Increases in the Services Sector

	Percent Average Annual Improvement	
	1960–1983	1970–1983
Telephone/communications	6.1	6.8
Air transportation	5.8	4.5
Railroad (revenue traffic)	5.1	4.8
Gas, electrical utilities	2.7	1.0[a]
Commercial banking	—	0.9[b]
Hotels/motels	1.6	0.8

[a] 1981 data.
[b] 1982 data.
SOURCE: Bureau of Labor Statistics, Office of Productivity and Technology.

create an ever higher level of per capita income. Part of this fear is the belief that services do not lend themselves to productivity increases through technology infusions. Our analysis suggests that the overwhelming proportion of productivity increases in both manufacturing and services have derived from capital and technological infusions. Between 1975 and 1982 there was a 97 percent increase in new technology investment per service worker (Office of the U.S. Trade Representative, 1983, p. 24). Some service industries have undergone significant improvements in productivity over the last two decades. In others, such improvements have not been very impressive (see Table 2).

At the margin, services and manufacturing seem about equally attractive to capital (see Figure 8). Consequently, one would expect a continuing willingness to invest in services for productivity improvements and competitive advantage whenever services can offer adequate returns. Because less attention has been given to automating services than to manufacturing, many opportunities to improve productivity still exist, and office automation is high on both producers' and users' lists of priorities (Collier, 1983).

Services can create real growth in per capita income as long as any of three conditions obtain: (1) the product sectors can produce enough output at continuously lower relative costs to release purchasing power for other desired (services) uses, (2) it is possible to increase productivity in existing services, or (3) entrepreneurs can conceive of new services having higher marginal value to buyers than existing services or products. Within wide limits, the services economy is a natural outgrowth of productivity increases in the goods-producing sector. Whereas agriculture once demanded some 70 percent of all employment in the United States, less than 4 percent of the work force now produces much more food per capita—including 50 percent more of the major grains than the country can eat (see Figure 1).

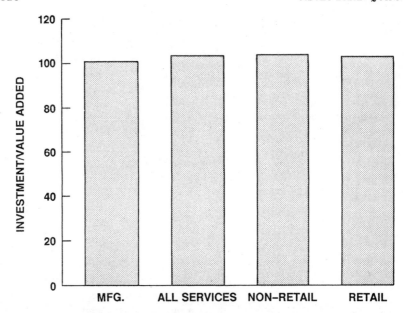

FIGURE 8   PIMS indices of investment efficiency. From PIMS 1985 data base, Strategic Planning Institute, Cambridge, Massachusetts.

A similar productivity phenomenon is now at work in manufacturing. With a decrease in the number of hours of work needed to produce or buy a basic automobile, radio, or washing machine, the percentage of the economy able to be devoted to other things naturally goes up. Since the average person can eat only so many pounds of food or use so many cars, washing machines, or appliances, the marginal utility of other things rises, and in recent years these things have often been services. As the perceived value (relative utility) of these activities increases, a given expenditure on services will create greater value at the margin and wealth is enhanced. For decades, the services sector has provided the U.S. engine for growth (see Table 3).

Once survival needs are met, the relative value (or utility) of all other objects, concepts, or services is solely a creation of the human mind. Pearls, gourmet foods, more comfortable furniture, vacations, phonograph records, parks, high-powered cars, or travel have value only because people create this value in their minds. Thus, there is no intrinsic limit to the wealth a services economy can create, other than the limits of human imagination in finding or placing higher values on new services. Such limits seem remote at the moment.

TABLE 3   Services Growth

Sector	Compound Annual Growth (percent)	
	1960–1984	1980–1984
Goods-producing	2.8	2.4
Agriculture, forestry, fisheries	1.4	3.1
Construction	0.7	1.2
Manufacturing	3.5	2.6
Mining	2.1	1.0
Services-producing	4.0	3.6
Communication	7.1	4.7
Telephone, telegraph	7.6	4.7
Radio, television	2.9	6.2
Finance, insurance, real estate	4.0	2.9
Banking	3.8	1.7
Insurance	3.0	0.2
Securities, commodities brokers	4.6	16.9
Real estate	4.2	3.3
Public utilities	3.9	1.9
Services	4.1	2.4
Amusements, recreation	3.4	4.9
Auto repair	4.2	1.5
Business services	6.8	7.2
Health services	5.2	3.6
Legal services	4.6	6.4
Personal services	0.4	1.9
Transportation	2.1	−1.0
Wholesale, retail trade	3.9	4.7

SOURCE: *New York Times* (October 27, 1985).

THE LIMITS OF CURRENT DATA

To help show how technology affects the generation of wealth through the services sector and the structure and competitiveness of a modern economy, this paper draws upon information collected from four data bases: the Bureau of Labor Statistics, the U.S. Department of Commerce, the Standard & Poors Corporation's Compustat Tapes, and the PIMS data base. Aggregate data for the services sector are compared against aggregate GNP and manufacturing sector data and are analyzed for each major services sector to reveal certain key variables and make appropriate comparisons among the sectors. This chapter also draws upon information obtained from knowledgeable technology-using concerns and from experts in five sample service industries: financial services, communications, public transportation (airlines and rental cars), health care and delivery, and retailing.

The information from these various sources makes clear that, although current data are replete with definitional problems and artifacts, the impact of technology on the services sector is powerful, misunderstood, and crucial to the future development of the United States and other advanced economies. This topic is worthy of much further study. This chapter, therefore, is not intended as a research report, but rather to help provide some initial insights on this complex issue.

Some Data Anomalies

Data anomalies abound. Most serious are those in defining what is included in measures of "the services sector." Specialized services (like product design or development, market research, accounting or data analyses) are captured as "manufacturing" costs if performed within manufacturing concerns, but as "services" if provided externally. Internal salesmen are reflected in manufacturing employment, but external sales representatives and wholesalers are services. If a farmer harvests his own grain, the costs become production costs; if the farmer hires a professional combine operator, the activity is a service. Home cooking and clothes washing are not measured as services activities, but restaurants and automatic laundries are. The oil that provides a home's heat is a product sale; but electricity for the same purpose may come from a "services sector" utility. And so on.

Productivity measurements pose a special problem (Mark, 1982). Although it is easy to count the number of autos or washing machines produced, how does one measure the output of a bank, insurance company, legal firm, or consultant in order to make productivity measurements? The output of a government agency is often measured only by the agency's cost (*New York Times*, October 27, 1985). One medical procedure may require fewer resources than another, but the (unmeasured) pain and morbidity it entails could be much higher. Many people may actually place more value on noncurative treatments or hope-creating reassurances (such as laetril or counseling) than a statistically beneficial procedure (such as chemotherapy). And so on again.

Defining the services component of international trade movements is even more complicated. If a loan officer in the British subsidiary of a U.S. bank relies heavily on analyses or expertise developed in New York for a deal he consummates in London, is this an export, a return on investment abroad, or just a local sale of services? Similarly, in manufacturing it is often impossible to say how much of a company's overseas profit is due to the company's products versus the embodied technology, training, systems, or services support the company provides indirectly to host country users from its U.S. base. How a cost or benefit is classified often depends on how the company is organized and the purpose of the measurement itself.

Such definitional problems are magnified by the ease with which services can be concealed in transfer pricing, cost allocation, or tax avoidance choices. Finally, trade in services is seriously distorted by the not too subtle trade barriers individual countries have placed on such trade, through their regulatory or monopolistic (banking, insurance, travel, postal, telegraph, and so on) home market restrictions (Sellers, 1985; *The Economist*, July 6, 1985). These will be attacked seriously for the first time in the next round of the Multinational Trade Negotiations.

Product Versus Service Interchangeability

Another artifact is the distinction (and preference) often attributed to production of products as opposed to services. This viewpoint ignores the truism that products and services are often interchangeable over a broad spectrum. Theodore Levitt, editor of the *Harvard Business Review*, has said, "A man doesn't buy a ¼-inch drill, he buys the expectation of a ¼-inch hole." A customer rarely cares whether a computer manufacturer accomplishes a function by a hardware circuit or internal software. Cable network services replace home antenna sales and provide more of what the customer is really buying, increased quality in reception and choice in programs. Computer-aided design and manufacturing software substitutes for added production machinery. Conversely, tree shakers and automatic harvesters substitute for teams of migrant pickers. Disposals and compactors lower garbage disposal costs. Home hair sprays replace beauticians. Gourmet processed foods and microwave ovens substitute for restaurants. And so on.

All of these substitutions create productivity and value-added increases that are just as real as the substitution of a machine for a factory worker or a new vacuum cleaner for a carpet sweeper. One set should not be denigrated and the other praised.

CHANGES IN INDUSTRY STRUCTURE

Recognizing these anomalies in the data, what useful observations or hypotheses can we make concerning the impact of technology on the services sector? What are some of the resulting effects on the overall structure of economic and competitive activities? Many potential effects are not yet realized because major technology introductions in several key services areas are relatively recent. Other effects are difficult to separate from those caused by the simultaneous deregulation of such important activities as airlines, communications, financial services, or common carrier services.

Two classes of technology affect service industries: industry-specific technologies (such as aeronautics for airlines or internal-imaging technologies for health care diagnostics) and generic technologies (such as communication, information-handling, data-storage, or transportation technologies) available to all sectors. Their effects on the service industries and overall economic competition differ somewhat. But their main impact is to allow significant changes in (1) economies of scale, (2) economies of scope, (3) output complexity, (4) functional competition, (5) international competitiveness, and (6) distribution of wealth. Such impacts occur virtually everywhere in the economy. Introductions of technology have created entirely new competitive situations for the service industries, their suppliers, and their customers domestically and internationally. They have changed the nature of manufacturing competition, and they have posed new technological, management, and policy opportunities and threats throughout the world.

Economies of Scale

Many service industries reported new economies of scale made possible by recently introduced technologies. The first-order effect in most cases was a new competitive structure characterized by both increased concentration and increased fragmentation (niching or segmentation). To obtain the full economies of scale available, large services companies merged into giant companies (*Standard & Poor's Industry Survey,* December 13, 1984). This phenomenon offers some interesting international trade policy opportunities. Since many services are inexpensive to transport internationally, nations or enterprises that achieve economies of scale early should enjoy some initial trade advantages externally and some entry barriers to their home markets. And countries whose economic policies permit such scale should benefit.

In the United States, new consortia have formed to provide new services no existing entity could handle alone. Many intermediate-size companies, unable to afford the new technologies, sold out to their larger brethren (see Table 4). Concentration measures in banking, transportation, financial services, and (less so) retailing all show increases from 1975 to 1985 (see Table 5). The communications industry was so affected by the AT&T

breakup as to make comparisons meaningless. After an initial concentration toward larger hospitals and delivery units, Diagnostic Related Groups and other economic considerations have begun to stimulate decentralization in certain aspects of health care. In all these industries, however, many smaller enterprises also identified local niches or specialized services and concentrated successfully on these. The result seems to parallel the classic "V"-shaped return pattern that Booz Allen & Hamilton, Inc., has reported for other industries, with higher returns accruing both to a few large entities and to those who specialize. For example:

- In the mid-1960s and early 1970s, automation of the securities-trading process changed the entire structure of that industry. Under the old market system, shares traded had to be physically delivered from the seller's agent to the buyer's. As daily volumes hit 10-12 million shares, only the big banks could hire and manage enough people to keep track of and clear their securities trades each day. Smaller firms began to fail because they could not control or process their securities. Finally, Wall Street firms joined together to form the Central Certificate Services (later the Deposit Trust Company) to immobilize virtually all the securities certificates under one roof. Then, rather than moving shares, a single set of accounting entries could control ownership. After 5-6 years the system became all electronic, and smaller brokers could tie into the depository. Now more than 175 million private and 175 million government transactions are handled yearly by automated clearinghouses (Office of Technology Assessment, 1984, p. 101).

  Later a group of New York banks formed the Clearinghouse for International Payments System (CHIPS), which handles transfers worth $60 trillion annually with same-day settlements (American Bankers Association, 1984). Trades can be made instantly by investment bankers all over the world, broadening and decentralizing the market vastly, despite the procedures required by the centralization clearinghouse. And specialized networks like Nasdaq, Intex, and Instinet now serve smaller, segmented, or localized markets (*The Economist*, July 6, 1985).
- Because of the reimbursement system then in use, the first effect of capital-intensive technologies in medical care was to centralize treatment into the hospitals. Small practitioners and hospitals did not have the patient volumes to afford the elaborate diagnostic, treatment, surgical, and recovery equipment that was being developed. This concentration allowed highly specialized medical practices to form around the hospitals and created even more specialized regional referral centers

TABLE 4 Merger Transactions

Industry Classification of Seller	Number of Transactions (rank) 1984	Four-Year Cumulative	Industry Classification of Seller	$ Millions Paid (rank) 1984
Banking and finance	251 (1)	1,343	Oil and gas	$42,981.8 (1)
Wholesale and distribution	143 (2)	498	Food processing	7,094.8 (2)
Miscellaneous Services	138 (3)	506	Conglomerate	6,982.9 (3)
Retail	130 (4)	393	Retail	6,673.2 (4)
Computer software, supplies, and services	118 (5)	399	Banking and finance	5,846.3 (5)
Oil and gas	102 (6)	369	Computer software, supplies and services	3,766.4 (6)
Industrial and farm equipment and machinery	93 (7)	379	Packaging and containers	3,089.4 (7)
Brokerage, investment and management consulting services	86 (8)	251	Insurance	3,005.9 (8)
Health services	86 (8)	244	Printing and publishing	2,863.9 (9)

Broadcasting	83 (10)	237	Chemicals, paints, and coatings	2,629.9 (10)
Chemicals, paints, and coatings	79 (11)	251	Leisure and entertainment	2,580.7 (11)
Instruments and photographic equipment	71 (12)	237	Energy services	2,546.2 (12)
Printing and publishing	67 (13)	225	Miscellaneous services	2,323.9 (13)
Electronics	67 (13)	271	Timber and forest products	2,297.2 (14)
Insurance	66 (15)	303	Electrical equipment	1,978.4 (15)
Drugs, medical supplies, and equipment	61 (16)	260	Broadcasting	1,917.9 (16)
Food processing	58 (17)	253	Health services	1,687.9 (17)
Office equipment and computer hardware	55 (18)	167	Wholesale and distribution	1,651.8 (18)
Electrical equipment	50 (19)	196	Industrial and farm machinery and equipment	1,635.6 (19)
Primary metal processing	43 (20)	141	Brokerage, investment, and management consulting services	1,460.3 (20)

SOURCE: Grimm (1984), pp. 40–41.

TABLE 5   Concentration Measures in Five Service Industries, 1975 and 1985

	1975	1985
Commercial banking		
Total assets	$525 billion	$1,018 billion
4 firm measure	0.36	0.42
8 firm measure	0.52	0.63
Life insurance		
Total assets	$213 billion	$526 billion
4 firm measure	0.47	0.43
8 firm measure	0.63	0.60
Diversified financial services		
Total assets	$127 billion	$603 billion
4 firm measure	0.29	0.40
8 firm measure	0.44	0.60
Retail sales		
Total sales	$112 billion	$297 billion
4 firm measure	0.31	0.32
8 firm measure	0.49	0.48
Transportation		
Total operating revenue	$36 billion	$107 billion
4 firm measure	0.25	0.30
8 firm measure	0.42	0.51

NOTE: Concentration measures represent the proportion of assets, sales, or operating revenues accounted for by the largest 4 or 8 companies in the industry. Totals represent summation of assets, sales, or operating revenues for the largest 50 companies in the industry. Some totals are estimated.
SOURCE: Compiled from Fortune Service 500 data, *Fortune*, July 1975, and *Fortune*, June 10, 1985.

to handle particularly difficult cases. The first-order effects were that many smaller hospitals suffered, closed down, or joined cooperative networks with the larger centers.

However, as uncontrolled costs soared (the average cost of an in-patient stay rose from $729 in 1972 to $2,898 in 1984) and patient care became more impersonal, new distribution systems emerged (U.S. Department of Commerce, 1985). Now medical services are integrating both vertically (home care, primary care, and specialty care) and horizontally (pediatrics, obstetrics, dermatology, internal medicine) into complex systems (such as health maintenance organizations [HMOs]) linked by electronic technologies. HMOs have grown in number from 39 in 1971 to 326 today and now serve more than 13.6 million people, compared with 3.1 million in 1971. The number of ambulatory surgical centers exploded from 400 in 1982 to 1,200 in 1985; and home health care was offered by 42 percent of U.S. hospitals in 1984, a 17 percent increase over 1983 (DeYoung, 1985). Management of these complex service systems has become such a critical factor that large private companies have found it profitable to apply their skills to hospitals and integrated health care management. To obtain specific economies, insurance companies have integrated forward into health care management, and some hospitals have begun to offer insurance (*Business and Health*, 1986a,b).

Economies of Scope

The introduction of new technologies has often created a powerful second-order effect—economies of scope—the capacity to provide entirely new service products through the same service network. This was especially true when the driving technology was electronic communications or information handling. When properly installed, such technologies often allowed their users to undertake a much wider set of customer, data, or services activities without significant cost increases—or even with cost benefits through allocating equipment, development, or software costs over a richer base of applications. In addition, new technologies frequently offered increased strategic potentials through timing advantages in introducing new products or fast response capabilities in dealing with competitors' moves. Such strategic flexibility can be the most significant payoff for companies in the services sector.

- In the mid-1960s, when the insurance industry was stable and heavily regulated, insurance companies automated their back-room activities to obtain dramatic gains in productivity in handling premium billings and collections.[4] As wildly fluctuating interest rates hit the industry in the mid-1970s, companies had to change their products rapidly to attract premiums and to offset the effects of customers borrowing against their policies at low interest rates. Only those companies that had flexibly designed computer and control systems could deploy their products rapidly enough to obtain a competitive edge. Persistence rates (losses or changes in policies over a 5-year period) jumped from around 5 percent to 30-40 percent. Insurance companies used to bring out new rate books every 3-5 years. Now both rates and products were presented in the electronics technologies. Industry executives said they could have neither conceived of the variety of new products needed nor explained and introduced them through their agents in a timely way without effective electronic and software systems. Smaller companies could not afford the huge initial costs of needed technologies and sold out or merged with larger companies who could benefit from their distribution networks. A flexibly automated back room became a key element in survival and competitive success.

Again, smaller or local groups had to create new relationships with those who had requisite technical capabilities in order to obtain new products in a timely fashion. Or they had to concentrate on localized or specialized services the larger company could not yet reproduce. The flexible potentials of telecommunications have led to an unprecedented series of national and international coalitions among large and small companies. Because of the need for compatibility among telecommunications systems, the formation of these coalitions perhaps led to (1) more rapid technology dissemination than ever before and (2) the obliteration of many potential national comparative advantages in these technologies. The worldwide affiliations of IBM and AT&T offer major examples (see Figure 9).

Two other industries suggest the basic restructurings caused by technology in the service industries:

- Recognizing that the capability for sales of financial services is now embodied in the electronic display system, Sears, K mart, and J. C. Penney have experimented in the $1.6 trillion retail financial services

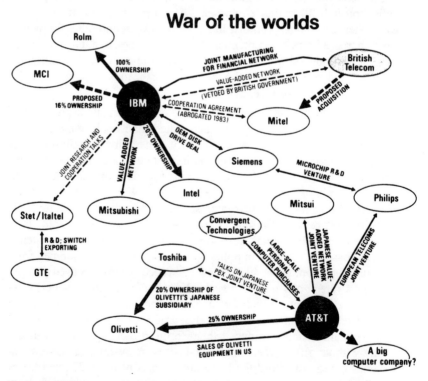

War of the worlds

FIGURE 9   When worlds collide. Reprinted with permission from *The Economist*, June 29, 1985.

market. Starting with its access to 40 million customers annually through its credit records, Sears now offers insurance (Allstate Insurance Co.), investment (Dean Witter Financial Services, Inc.), and full real estate services to customers in many locations (Gardiner, 1985). Through its savings bank, credit card, and automatic teller machines (ATMs), it is now extending other financial services to its most remote branches. Although the nation's 45,000 ATMs (U.S. Department of Commerce, 1985) already offer highly decentralized access to financial services, more than half are hooked into one of seven national networks such as Citishare or Cirrus (*United States Banker*, August 1985, p. 10). Many observers expect home banking services and electronic funds transfers (EFTs) to extend these services even further. Whereas only 31 percent of transactions are made through EFT today, Arthur D. Little estimates this represents 90 percent of the total value of all transactions. McKinsey & Co. has forecast a three-tiered future banking structure in which about 10 commercial banks and a few other financial institutions will operate on a national level (Cooper and Fraser, 1984, p. 216). A few new entrants will use technology to bring specialized services (such as clearing centers or discount brokerage) to the middle tier, and the regionals and local banks will scramble to find coalitions or service highly specialized local needs.

- In rental cars, the large companies in conjunction with airlines can lock up the general or business traveler with instantaneous guaranteed reservations virtually anywhere. All have international subsidiaries or affiliates connected electronically. But niched operators have proliferated under their pricing umbrella. Automatic telephone answering devices allow both the super elegant and the Rent-A-Wreck extremes to serve local markets with office-in-the-home scales of operations. Alamo and Agency rental companies have segmented target groups with specific needs.

Output Complexity

Technology in services permits complexity. Engineering specialists, molecular biologists, or epidemiologists can analyze and resolve more complex problems using computer models and data base networks than ever before. The effects are so great that in some research areas, Joshua Lederberg, president of the Rockefeller University, suggests the technology is moving from "information search to knowledge search," that is, computer technologies can identify relationships and pose new hypotheses as well as merely analyze and test data. For example, computerized analytical models can suggest entirely new structures for complex proteins, and gene machines can manufacture them within hours for testing against potential antigen or biological insults. Entertainment and educational media can achieve effects never before attainable. Law firms can accomplish more complex searches of legal background, prepare more intricate contracts, and negotiate and document the resulting settlements more thoroughly in hours than they previously could in weeks. The velocity of transactions in monetary exchanges and in commercial activities has increased so greatly that firms cannot compete without well-developed electronic support systems. These shorter transaction times have led in turn to potential economic volatilities that can threaten the capacity of even great companies or nations to control their economic destinies.

The complexities that service technologies allow have created some bizarre twists integrating world economies in unexpected ways.

- Once the financial houses could process documents efficiently, they began to look for ways to market effectively and to sell products off their technology bases. Among other things, a proliferation of products for pension funds appeared, including master trusts, securities lending, and international custody. GTE Corporation now has 40 different firms managing its pension fund worldwide, yet these investments can be monitored, handled, and settled through a master trust anywhere in the world every day of the week—impossible without electronics technologies. With this capacity, large pension funds (like the $17 billion fund of General Motors Corporation) can make or lose enormous amounts of money on small changes in stock prices or interest rates.

  Because these funds can trade large blocks quickly, the volume of trades has increased by orders of magnitude in the last decade. Daily volumes traded on the New York Stock Exchange grew from 10-12 million shares in the early 1970s to more than 102 million in 1984, with more than half of the trades involving 10,000 or more shares (*Wall Street Journal*, April 22, 1985, p. 1). Since the pension funds with large holdings of common stocks could trade profitably on small point spreads, the secondary consequence of this new technology has been tremendous pressure on short-term profit and stock price performance. This pressure has plagued all U.S. companies—manufacturing or otherwise—trying to increase productivity through long-term investments in technology. As financial markets further internationalize, this pressure may extend equally to all nations' publicly held companies.

The technologies that allow manufacturers to produce a higher variety and quality of products at ever-lower costs also open entirely new market niches requiring sophisticated new marketing and services practices never needed before. But computer-related technologies now permit retail establishments to deal with this greater range of offerings at a lower cost and under better control than ever before. Again, technology creates a higher service value (or utility, hence real growth) for consumers by matching their more complex tastes and localized needs more explicitly— and usually at lower cost. For example:

- Computer technology allows rental car companies to analyze their costs for each type of customer. Because these costs vary enormously, the result has been an elaborate set of rate structures by city, length of rental, day of week, season, holiday pattern, and so on. Interestingly, if customers cannot figure prices accurately, they concentrate more on services rendered, allowing higher margins to those who can satisfy their needs. Computerized airline reservation systems allow similar complexity, but they also permit unique service values to customers, such as: specialized meals, wheelchairs, committed seat assignments, luggage verification, and even customer counseling for nervous or new flyers. Bar-code scanners in retailing allow retailers instant feedback on their sales and inventory movements—and hence the capacity for much more complex inventory and cost management systems (*Chain Store Age Executive*, January 1985, p. 3). Some are now selling this information back to market research firms who use it to help manufacturers fine-tune their strategies. Sophisticated information systems are allowing major chains to branch out and manage a portfolio of retail drug, supermarket, home center, and small specialty chains with much higher value added potentials (*Chain Store Age Executive*, March 1985, pp. 25-27).

Unfortunately, to use the full sophistication of modern technologies, data about individuals' health, wealth, and activities are stored in many accessible places. And disastrous data errors, frauds, or system failures are possible. To protect individuals and institutions in this environment will require further sophisticated developments in technologies, institutional attitudes, professional standards, and national and international regulations. The President's Office of Telecommunications Policy has suggested some of the likely key issues for public policy (National Science Foundation, 1975; Rule, 1975).

- Medical technologies perhaps provide the best example supporting the complexity argument. They can now diagnose, attack, and resolve most of the common infectious diseases and surgical problems of the past, switching the medical care system's attention toward the prevention and cure of much more complex diseases and morbidity patterns. Mortality rates have shifted markedly (see Table 6). But as

TABLE 6   Causes of Death per 100,000 in the United States

	1900	1978
Influenza/pneumonia	202	27
Tuberculosis	194	1
Gastroenteritis	142	0
Nephritis	81	0
Diptheria	40	0
Cardiovascular	137	443
Malignancies	64	182

SOURCE: National Center for Health Statistics (1980).

technology improved, so did people's expectations and use of the system. Whereas medicine offered few nonsurgical "cures" (other than arsenic for syphilis and quinine for malaria) until the antibiotic revolution of the 1930s, by the late 1960s patients began to *expect* cures as a routine matter. They began to use the system for a much wider and more complex set of purposes and began to sue if their expectations were not met, regardless of the probabilistic or medical realities. Medical technologies in many cases became so powerful that life, according to some definitions, could be sustained almost indefinitely. This has, of course, led to still another level of technical, legal, and ethical complexities.

Functional or Cross Competition

With such diversity has come the breakdown of traditional industry demarcations and significantly increased functional or cross competition. These changes are most obvious in financial services.

- Disintermediation occurred because consumers found they could communicate effectively directly with the market and no longer cared what—if any—intermediary came between them and their investments. Banks, insurance companies, brokerage houses all began to offer a similar range of financial products and services (see Table 7).

  Soon retailers and manufacturers (such as GM and Ford) used their credit bases to present comparable offerings. The distinction between financing and product sales rapidly disintegrated—note the success of 2.9 percent financing campaigns for automobiles. General Motors Acceptance Corporation's $54 billion in assets and $1 billion in profits (more than the corporation's total profits in 1983) make it the nation's largest single holder of consumer debt (Gardiner, 1985, p. 9) and suggests the potential impact of such activities on manufacturing firms. In addition, financial services became another way to attract customers into retail stores—for example, Sears' use of Allstate and Dean Witter on the one hand and First Nationwide Financial Corporation's joint venture with K mart Corporation to deliver its services on the other.

- Communications, electronics, and printing technologies have allowed national and international newspapers to attack the advertising base of magazines and television networks. Communications technology has enabled airlines, rental car companies, and hotel chains to join together to offer complete vacation packages that were once solely the purview of travel agents and tour groups. Courier services have essentially integrated private airlines and ground services to compete selectively with telephone, facsimile, telegraph, and mail services. Through their electronically managed "wholesale clubs," retailers have integrated backward into wholesaling and into some manufacturing.

One upshot of these changes is more rapid introduction and delivery of products or services with worldwide-scale economies and quality into the most remote markets of the United States and other advanced countries. In such countries, virtually all competition now has international dimensions. A second result is that consumers' attempts to enjoy both "one-stop" shopping and competitive prices create new possibilities for inventory and distribution economies. Major manufacturers can integrate completely from "just-in-time" suppliers to the shopping mall—as GM seems to intend with its Saturn automobile project. Although potential gains appear high, with such integration comes increased risks—notably GM's $5 billion

TABLE 7   Financial Services Offered by Depository and Nondepository Institutions

	Banks	S&L's	Insurers	Retailers	Securities
Checking	a,b	b	b	b	b
Saving	a,b	a,b	b	b	b
Time deposits	a,b	a,b	b	b	b
Installment loans	a,b	b	b	b	b
Business loans	a,b	b	b	b	b
Mortgage loans	b	a,b	b	b	b
Credit cards	b	b	b	a,b	b
Insurance			a,b	b	b
Stocks, bonds			b	b	a,b
Mutual funds			b	b	a,b
Real estate			b	b	b
Interstate facilities			b	b	b

[a] 1960.
[b] 1982.
SOURCE: Koch and Steinhauser (1982).

investment in Saturn—if a mistake is made. Conversely, if competitors miss this strategic opportunity, they could suffer major losses.

So pervasive are communications and electronics technologies that a key element in competitive strategy for even "product-oriented" companies such as Exxon or GM is information management. Their profits can be made or broken by how well they develop and deploy knowledge about supply costs, new technologies, exchange rates, changing regulations, swap potentials, political or market sensitivities, and so on anywhere in the world. In the aggregate, more money may be made in the goods sector through information and services than "production" activities. The U.S. Trade Representative's Office in its U.S. National Study of Trade in Services (1983, p. 13) cites an estimate that about three-fourths (or 25 of the 33 percent) of the total value added in the "goods sector" is created by services activities within the sector.[5] If this level of contribution is correct, one can expect to see information managers and similar services technologists increasingly rising to the top of major producing organizations—as they already have in oil companies, banks, and the retail trades. The cross substitution between production and services will become ever more apparent.

International Competitiveness

Perhaps the most perplexing, yet crucial, impact of services technologies will be on international trade. Within the definitional limits cited, *The Economist* (October 12, 1985) estimates only 18 percent of world trade to be in services as opposed to 49 percent in manufactures and 33 percent

TABLE 8   World Comparisons

	Average Annual Compound Growth Rate, 1970–1980 (percent)	Value in 1980 ($U.S. billions)[a]
Services exports[b]	18.7	350
Merchandise exports	20.4	1,650
Foreign investment income[c]	22.4	225
World production	14.2	9,389

[a] Converted from SDRs and nominal values in national currencies at current exchange rates to U.S. dollars. World is defined as IMF member countries reporting data for both 1970 and 1980.

[b] Services exports exclude official transactions and investment earnings.

[c] Foreign investment income includes private direct investment income and portfolio income but excludes official transactions.

SOURCE: Derived from various issues of *Balance of Payments Statistics*, International Monetary Fund, and *International Financial Statistics*, International Monetary Fund.

TABLE 9 Ten Largest Services Exporters 1980 (billions of U.S. $)

Country	Value of Services Exports[a]	Value Foreign Investments Income[b]	Services Balance	Services Exports— GDP (%)	Services Exports— Merch. Exp. (%)
United States	34.9	70.2	6.0	1.4	15.6
United Kingdom	34.2	17.1	9.8	6.5	30.9
France	33.0	18.4	5.5	5.1	30.7
Germany	31.9	8.5	−17.9	3.9	17.2
Italy	22.4	5.3	6.2	5.7	0.2
Japan	18.9	7.2	−13.4	1.8	14.9
Netherlands	17.7	10.0	0.2	10.5	26.2
Belgium	14.5	17.6	0.5	12.1	26.3
Spain	11.7	0.2	6.3	5.6	56.9
Austria	10.8	2.5	5.1	14.0	62.6

[a]Services exports exclude official transaction and investment earnings.
[b]Foreign investment income includes private direct investment income and portfolio income but excludes official transactions.
SOURCE: Derived from various issues of *Balance of Payments Statistics*, International Monetary Fund, and *International Financial Statistics*, International Monetary Fund.

in other goods support activities. Admitting that the $100 billion statistical discrepancy in world trade balances is probably due to a substantial underestimation of services trade, the U.S. Trade Representative's Office estimates $350 billion in world services trade in 1980 as opposed to $1,650 billion in merchandise trade (see Table 8). Yet even these figures are misleading because so many of today's sophisticated products could not be sold abroad without supporting services to finance, maintain, and upgrade them in the marketplace.

Worldwide services exports in 1980 were only about 3.7 percent of gross domestic product, and those of the United States were only 1.4 percent, but worldwide services trade was growing at a 19 percent rate versus merchandise's 20 percent (Office of the U.S. Trade Representative, 1983, p. 13). Both rates were much greater than world production's 14 percent growth. Twenty-four countries accounted for 87 percent of services exports in 1980, and the United States led with $34.9 billion (see Table 9).

Technologies related to services have vastly restructured the international marketplace in other important ways. Communications technologies, of course, permit manufacturers to coordinate their design, sourcing, distribution, and manufacturing activities worldwide to minimize costs. But economies of scale in services affect manufacturers' scale economies in other ways as well.

• Containerization and new on-board storage techniques (such as liquefied natural gas) facilitate and lower the cost of trade for such things as expensive high-technology goods, volatile chemicals, and coal. This trade in turn has led to major ($250 billion in 1981) construction and engineering projects in both buyer and seller countries that require more traded services and the development of technical support (services) industries in both exporting and importing countries. Cheaper and more flexible transportation systems also have lessened geographic constraints on production and further encouraged transfers of skills, technology, and knowledge among countries (Office of the U.S. Trade Representative, 1983, p. 16).

- Large-scale investor arrangements now allow direct access to the Eurobond market where large "blue chip" companies can buy Eurobonds directly (often at lower interest rates than U.S. treasury bonds) and can place other securities directly with large investors or on a "bought deal" basis, thus lowering their capital costs relative to smaller companies. The extent of these transactions has become significant. By 1985 the market for Eurobonds denominated in dollars was $38.4 billion, and the commercial paper market was more than $230 billion (*The Economist*, March 16, 1985). An international "swaps" market of more than $20 billion had also developed (Credit Suisse-First Boston Bank, 1983) enabling two sophisticated companies to borrow in the domestic markets that were most favorable to their needs (interest levels or rate stability), then swap their interest obligations, splitting possible interest savings.

World capital markets have become so integrated that it is difficult for any single nation's producers to achieve a capital cost advantage over other international competitors. Local banks or large companies can trade directly in virtually any money market in the world, as clearances are made instantaneously by electronics. At present there is continuous trading on some major market during all but 6 hours each day. Soon trading will be a 24-hour-per-day phenomenon. The business in foreign exchange transactions is already $6 trillion annually, and it is expanding. Under these conditions, the extent to which sovereign nations can intervene effectively to control their monetary systems (and hence inflation rates) in order to manage their economies through traditional means is not clear. This may be the most significant single impact of technology in the services sector.

- In 1985 computerized quotations and satellites had global financial markets; more than 500 companies were listed on at least one stock exchange outside their home countries. Innovations in one market were rapidly reflected in others. The European "bought deal" procedure led the U.S. Securities and Exchange Commission to issue SEC Rule

415 allowing companies to preregister issues and sell them off-the-shelf as needs or market conditions warranted. The negotiated rates of U.S. brokers are slowly being forced onto the exchanges of other countries. The United States has relaxed requirements of its bond markets to make them more competitive with Eurobonds; and foreign exchanges (notably, London and Tokyo) have opened their memberships to foreign brokers. In investment banking, "bought deal" procedures have shrunk spreads and concentrated underwriting to the few firms that have the capital bases, distribution outlets, and trading systems to handle them. In 1984, two-thirds of all such new issues were handled by only five firms in the United States (*The Economist,* March 16, 1985).

One initial impact of improved data, production control, and communications technologies has been to allow—or force—manufacturing outside of industrialized (or wealthy) countries to move toward developing areas where costs of labor or materials have been lower. With advanced electronic technologies, overseas facilities could be controlled or managed with whatever degree of centralized coordination competition demanded. Although this use of technology has distributed manufacturing more widely, the longer-term effects of decentralized production are less clear. The diversity of specialized customer tastes that can be accommodated by flexible manufacturing systems and better communications technology may soon create higher competitive premiums for those who stay physically closer to the marketplace and can respond more rapidly to customers' needs for services.

- Saturn's planned 8-day cycle from order to delivery of an options-loaded vehicle may make it difficult for foreign producers to compete. The same may be true of just-in-time inventoried products throughout the U.S. production system. Japanese auto companies manufacturing in the United States are now bringing their suppliers here for just this reason. Interestingly, manufacturing may actually return to large advanced economies because of the "services" need to satisfy a highly diffuse marketplace rapidly and flexibly.

A crucial question is whether, without domestically controlled manufacturing and R&D, a country can have a healthy technological community. In the past, manufacturing enterprises have been the locus of most R&D both for their own consumption and for performance of government R&D contracts (see Figure 10). If U.S. companies' manufacturing plants were located overseas, U.S. R&D could support them for a while. But it is difficult to maintain manufacturing R&D capabilities without the close interaction and feedback a local plant offers.

As other nations' enterprises achieve world-scale economies with their

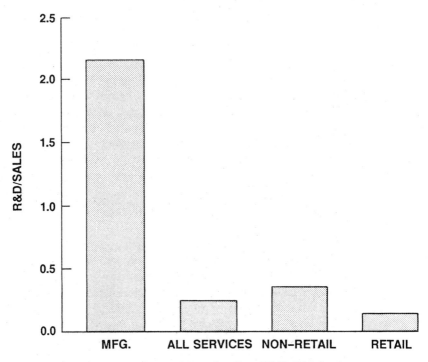

FIGURE 10   PIMS indices of research intensity. From PIMS 1985 data base,
Strategic Planning Institute, Cambridge, Massachusetts.

own domestic plants in maturing markets, it may be nearly impossible for
U.S. concerns to stay abreast in manufacturing without domestic plants.
Many competitors are sadly aware of how important user proximity (domestic plants and markets) has been for Japanese productivity. Although
users of services can provide a significant driving force for product technologies, simultaneous development of product and manufacturing capabilities has become so important that lacking the latter could be the fatal
flaw in a services economy. Both wealth and (more important) intellect
could leak away to other "producing" countries.

Nevertheless, services themselves are also a critical cost dimension of
a nation's competitiveness in international production. Internal transportation, communications, financing, health care, distribution, and other
services can lower the real cost of manufactures. Efficiencies in these
sectors also improve the real wealth of laborers for any given wage level;
that is, workers can buy more goods and services per dollar earned. For
example, the Japanese have been superb in certain mass-manufacturing

fields, but their productivity (in GNP per person) has consistently lagged behind that of the United States, largely because of a less-productive services sector (see Table 10).

As an economy moves ever closer to a total services base, a most important question emerges: What does the nation trade to obtain the manufactures and raw materials it needs from external sources? Although U.S. manufacturing and agricultural exports have suffered notable relative declines in recent years, the United States has had a strong positive net balance of trade in services and income from investments abroad, excluding payments for investments in the United States (see Table 11). Some have questioned whether traditional arguments of comparative advantage will be relevant in world services trade, especially in financial or information-based services where everyone can buy the same hardware (and often the same software) and connect into the same networks (Deardorff and Jones, 1985). This problem is compounded in the technology because the developers of much of the technology used in services trade are suppliers (often manufacturers) whose incentives are to sell and introduce their technologies as widely and quickly as possible worldwide.

With rapidly advancing generic technologies such as electronics and communications driving the services industries, it will be difficult to establish or maintain a national competitive advantage in any given services industry. Nations' trade and economic policies may have to focus more on improving education infrastructures and removing barriers to fast and flexible deployment of technologies and less on traditional investment-oriented industrial policies (Grossman and Shapiro, 1985). For example, *The Economist* (November 29, 1985) asserts that by early deregulation of

TABLE 10   Comparative Services Productivity, United States and Japan (dollar output per hour)

	Japan		U.S.		U.S./Japan
	1970	1980	1970	1980	1980
Private domestic business	3.59	6.01	9.40	10.06	1.67
Agriculture	1.37	2.38	16.53	18.36	7.71
Selected services					
Transportation and communication	3.86	5.66	9.29	13.14	2.32
Electricity, gas, water	14.01	19.74	21.98	25.38	1.29
Trade	2.88	4.53	6.88	7.92	1.75
Finance and insurance	6.69	12.03	8.21	8.20	.68
Business services	3.39	3.60	7.69	7.59	2.11
Manufacturing	3.91	8.00	7.92	10.17	1.27

SOURCE: UNIPUB (1984).

TABLE 11   United States Net Trade Balance^a (billions of current dollars)

Category	1965	1970	1975	1980	1981	1982	1983	1984
Net goods and services balance	8.3	5.6	22.8	9.0	13.2	0.1	−31.9	−90.1
Merchandise balance	5.0	2.6	8.9	−25.5	−28.0	−36.4	−62.0	−108.3
Services balance	0.2	0.2	1.8	6.3	8.3	7.4	4.8	0.8
Net investment income	5.3	6.2	12.8	30.4	34.1	29.1	25.4	19.1

^{a}Excludes military transactions.
SOURCE: *U.S. International Transactions 1960–1984*, Washington, D.C.: U.S. Department of Commerce, Bureau of Economic Analysis.

communications markets, the United States gained a lead in both use and production of communications technologies that Europe's more regulated sectors may never close. On the other hand, there is little room for complacency. Many countries and companies have proved that their skills in managing services enterprises are formidable indeed. The United States must work hard not to dissipate its lead in communications as it did in manufacturing. Ominously, however, many of the same causes of lost position are beginning to appear in this sector, namely, a short-term orientation, inattention to quality, and overemphasis on scale economies as opposed to customers' concerns.

The trade balance in merchandise has been negative in 11 of the last 24 years. In more than half of these, however, services (including investment returns) have put the current account in the black. But even adding $10 billion for the U.S. 10 percent share of understated world services trade (Office of the U.S. Trade Representative, 1983, p. 108), long-term prospects do not look encouraging for U.S. trade balances unless manufacturing returns to the United States and the strong U.S. dollar weakens. Table 12 gives a detailed breakdown of U.S. services trade.

Many experts believe that the net effect of services trade is even more seriously understated by definitional and reporting biases. For example:

- As their major customers increasingly sought international raw materials, supply sources, economies of operating scale, or markets, both U.S. and foreign banks followed their customers overseas in the 1970s. By the early 1980s, 30-40 percent of all U.S. bank profits came from international operations, with many of the money center banks exceeding 50 percent (*The Economist*, March 16, 1985).

The International Trade Commission cited an estimate that in 1982

TABLE 12 U.S. Detailed Services Trade Transactions (millions of dollars)

	1978	1979	1980	1981	1982	1983	1984ᵃ	Change, 1983–1984
Service transactions, net	23,625	32,241	34,487	41,129	35,327	28,143	16,986	−11,157
Receipts	77,940	102,323	118,216	138,636	138,250	131,944	142,010	10,066
Payments	−54,315	−70,082	−83,729	−97,507	−102,923	−103,801	−125,024	−21,223
Military transactions, netᵇ	621	−1,778	−2,237	−1,115	195	515	−1,635	−2,150
Travel and passenger fares, net	−2,585	−2,000	−825	58	−1,599	−5,064	−7,830	−2,766
Other transportation, net	−988	−935	−172	86	591	480	−976	−1,456
Fees and royalties, net	5,215	5,352	6,360	6,560	6,938	7,402	7,577	175
Investment income, net	20,565	31,218	30,443	34,052	27,803	23,508	18,115	−5,393
Direct, net	21,247	31,826	28,488	25,496	18,140	14,023	12,351	−1,672
Other private, net	6,149	8,173	11,905	21,629	23,641	22,310	20,425	−1,885
U.S. government, net	−6,831	−8,781	−9,950	−13,073	−13,978	−12,825	−14,661	−1,836
Other private and U.S. government, net	798	383	917	1,488	1,401	1,302	1,734	432
Contractor operations, net	1,348	1,054	1,591	2,027	2,398	1,790	2,109	319
Reinsurance, net	−532	−617	−624	−606	−590	−506	−553	−47
Communications, net	−65	−143	−317	−466	−758	−724	−721	3
U.S. government, net	−925	−1,198	−1,332	−1,366	−1,705	−1,563	−1,528	35
Other, net	972	1,287	1,599	1,900	2,057	2,306	2,427	121

ᵃPreliminary.
ᵇConsists of goods and services transferred under military sales contracts less imports of goods and services by U.S. defense agencies.
SOURCE: *Survey of Current Business*, March 1985, Table G.

nearly 25 percent of U.S. merchandise exports went to U.S. services businesses overseas. Some services, such as drilling, minerals exploration, civil engineering, banking, communications, or transportation, can be readily exported or are necessary purchases an outsider must make to trade in the economy. Certain technologies can be exported through licensing agreements. However, most such agreements pertain to manufacturing or product technologies. Without significant production inside the parent country (for example, the United States), a nation's ability to generate international services revenues through royalties or technology payments may be seriously impaired. How serious this impairment could be is unknown. Many experts believe that, unless manufacturing reverts to advanced countries through mechanisms like those suggested, increasingly services-oriented economies of advanced countries could lead to a serious and continuing weakening in their world trade positions, their strategic capabilities, and the value of their currency in world trade. What the net effect might be as all affluent countries move toward services economies needs serious research.

Growth and Distribution of Wealth

What are the effects of a services economy on distribution of growth and wealth, domestically and internationally? How does technology affect the process? For nonsupervisory workers, weekly average wages in manufacturing are about $373, whereas wages in services are about $250. This gap is overstated because more than 20 percent of all services workers are employed part-time (less than 30 hours per week); part-time workers constitute less than 5 percent of all manufacturing employees. Hourly wages in specific industries and current trends paint a more encouraging picture for services. Although average hourly wages per worker are higher in manufacturing than in some major services industries, notably retailing, other services activities enjoy higher average hourly wages than manufacturing, and the gap is closing in financial and other services (see Table 13).

Job opportunities in the United States have, of course, been growing most rapidly in the services sector for years. But recent job growth has been dramatic. From 1948 to 1978, manufacturing jobs grew by 3.6 million, but only 600,000 of these employees were in production jobs. More recently, 642,000 jobs were lost in manufacturing from February 1981 to February 1985, but services employment grew by 3.1 million. Since more affluent people spend a higher percentage of their income on services, this trend is likely to continue for the near future. The Bureau of Labor Statistics (BLS) also estimates that about half of all new manufacturing jobs created between 1969 and 1979 were white collar. Neal Rosenthal of BLS's Division of Occupational Outlook estimates that the

TABLE 13   Average Hourly Wages per Worker

	1983 ($)	1984 ($)	1983-1984 Growth Rates (%)
Average nonagricultural	8.02	8.33	3.9
Manufacturing	8.83	9.18	3.9
Durable	9.38	9.74	3.6
Nondurable	8.08	8.37	3.6
Transportation and utilities	10.80	11.11	3.2
Wholesale trades	8.54	8.96	4.7
Retail	5.74	5.88	2.6
Finance, insurance, real estate	7.29	7.62	4.5
Other services	7.30	7.64	4.3

SOURCE: *Survey of Current Business*, June 1985, Table S12.

shift to services employment in the last decade has actually decreased the percentage of workers holding low-paying jobs (Kirkland, 1985). BLS forecasts to 1990 suggest that low-paying services jobs will keep pace with, but not exceed, total growth in employment. But some service areas with high-paying jobs (such as computer services and investment banking) are expected to have high growth (see Figure 11). More than 60 percent of U.S. employment is now in the information industries, and virtually all of the 20 highest-growth occupations in the 1980s, as forecast by BLS, are in information handling. The Fishman-Davidson Center (University of Pennsylvania) showed that those states with the highest proportions of services employment also had the highest real income averages (see Figure 12). However, which is the cause and which is the effect is not clear.

Although some observers have suspected that the shift from manufacturing to services was a prime cause of the productivity slowdown in the United States in the 1970s, Kutcher and Mark (1983) found that such changes accounted for less than 0.1 percent of the change in productivity growth from 1959 to 1979. A real culprit, however, was the shift from high to low productivity goods-producing industries, accounting for up to 0.6 percent of the slowdown per year. Since many services jobs must be close to the point at which the services are used, services employment tends to become more geographically dispersed, following people's preferences for suburban and rural living. The quality of employment thus improves on two scales. Many physically difficult or hazardous jobs in production disappear in favor of "white collar" jobs in services, and the location of jobs is generally more pleasant and convenient. Services allow more part-time jobs for multiple-income families, and there is evidence that the family income for those employed in services may thus be higher than for those in manufacturing.

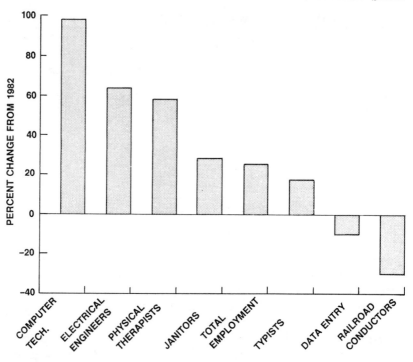

FIGURE 11   Projected job growth by 1995. From Bureau of Labor Statistics.

SOME ULTIMATE QUESTIONS

This discussion may lead to several ultimate questions about services economies. Can a services economy remain wealthier in the long run than more manufacturing-oriented economies? How far can an economy move toward a services base before it can no longer maintain its relative wealth? Can a services economy support the R&D necessary to maintain intellectual leadership and a high level of productivity growth?

Can Services Allow Greater Wealth?

As productivity increases in various manufacturing sectors, a large country such as the United States can reach self-sufficiency in a wide variety of products, employing only a small percentage of its population in manufacturing—just as less than 4 percent of the U.S. population now in agriculture produces a surfeit of food. Once reasonable self-sufficiency is obtained in a modest range of production, the definition of "wealthier" becomes more subjective. It depends on the relative value placed on different goods or services by the society. A more stable, safer, healthier society with fewer goods could be considered wealthier than one with more goods.

Although some observers claim that services industries are inherently incapable of creating "wealth" that can be transferred to future generations, even this argument fails. Better education, art, literature, health care, cultural capabilities, convenience in transportation, communication capabilities, recreational availability, and personal security can be transferred to future generations. These services have been the true measures of wealth throughout history. Thus, services societies can easily be wealthier than production-oriented economies—especially if the latter must pay a high premium in environmental degradation. In fact, some observers believe a services-driven economy may represent the most advanced level of economic development.

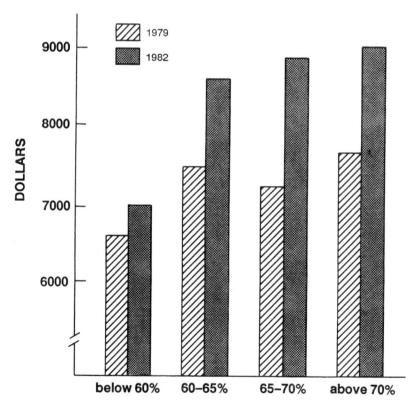

FIGURE 12   Average state real per capita income by percent of total private nonfarm employment in services in states. From Fishman-Davidson Center (1985).

What Is the Basis of World Power?

We often equate manufacturing power with economic wealth and world power. We forget that in the past the wealthy and great nations tended to be the trading nations, the educated nations, and the money centers of the world. Commerce, not manufacturing, led to wealth and power. Has the world changed fundamentally in the last 100 years? Perhaps so. Military power recently seems to depend on production power. But today selected intellectual and research capabilities in high technologies may be more important militarily than massive production potential, especially if a war is short. Nevertheless, it seems unlikely that a large modern country could maintain its strategic viability without world competitive aerospace, steel, chemicals, transportation, electronics, communications, and related support industries.

Some of these industries may need significant government support to be strong enough to meet the needs of defense, but fortunately most do sell extensively to the services sector. This sector could, if properly stimulated, provide the basic technological demands to maintain an important level of defense preparedness. Industries such as airlines, communications, information systems, financial services, and rental cars already hold such potential. Careful analysis is needed to explore the strategic issues raised by an economy increasingly oriented toward services.

OVERALL IMPACTS: TECHNOLOGY IN SERVICES

Technological advance is rapidly revolutionizing modern economies through services and presenting entirely new opportunities and challenges for corporate and national policymakers. The old "mom and pop store" and "hand laundry" analogies are anachronistic. Technology has created services industries of a scale, sophistication, complexity, and value-added potential to match those of any manufacturing industry. In fact, services and manufacturing are inextricably intertwined. Services industries are among the most important customers and suppliers for manufacturing. They buy many of manufacturing's highest-technology products and they provide important inputs for manufacturers, offering the latter opportunities to have lower cost than foreign competitors in critical areas. Services substitute so broadly and directly for manufacturing functions that no manufacturer's strategy is complete without a thorough consideration of how services (or services technologies) can contribute to the company's productivity, value added, growth, flexibility, and output quality.

In international trade, services create strong relationships between a foreign company and its host countries. Most of the benefits of services industries—the "product" as well as jobs and facilities—accrue to the host country, thus developing a strong mutuality of interests between parent companies and host countries. At present some U.S. services companies enjoy economies of scale and scope their international competitors cannot equal—except in banking and communications. And the deregulated U.S. marketplace provides them a unique stimulus for innovation. If U.S. services companies move aggressively to develop their own proprietary technology systems, they can maintain a 1- to 2-year competitive edge in most services areas. Any slowdown or delay in such innovations will be sure to attract competitive incursions in the U.S. and world services markets—as has already occurred with Japanese banking, tourist, hotel, and airline expansions and significant European acquisitions in U.S. distribution and tourist trade activities.

National sovereignty may be challenged in new and significant ways by the emergence of modern technologies in the services sector. An individual country will undoubtedly find it harder to control some of its most important resources; for example, information, monetary flows, and intellectual property. Interdependence and diffusion in these areas, however, could lead to greater world stability and less disparity among nations. As capital and information increasingly flow electronically across borders, a real question exists whether traditional comparative advantages are possible, in the long run, and if not whether the very basis for trade decreases. If nations or corporations cannot capture the benefits of their research, will they continue to perform it? Or will they be forced to even more frantic efforts to maintain at least a short-term edge that makes profits possible?

The long-term structural shift to services raises intellectual questions and important policy issues but, while there are certainly some problems, it seems inappropriate to be afraid of a greater services economy—or to deride it. A greater fear should be that nations misunderstand the services sector, underdevelop or mismanage it, and overlook its great opportunities while shoring up manufacturing industries at great national and corporate costs.

ACKNOWLEDGMENTS

The author gratefully acknowledges the generous contributions of Bell and Howell Company, Bankers Trust, and the Royal Bank of Canada in supporting the research for this chapter.

NOTES

1. Note that even here the product (clothing) has value only in relation to the service (protection) it provides its possessor.

2. Annual reports, Sears Roebuck and Co., and J. C. Penney, K mart, WALmart, and Zayre Corporations.
3. This conclusion is based on the PIMS data, which are self-reported by relatively large companies, and like other data bases on service industries have some inherent definitional problems.
4. London banks, for example, reported handling 128 percent more clearings in 1982 than in 1973 with only a 33 percent increase in personnel (*The Economist*, July 6, 1985).
5. Studies of productivity in manufacturing also indicate that a typical product is worked on during only about 8-10 percent of its production cycle; some 90 percent of its cycle is consumed in movement, waiting, inspection, and other support activities.

REFERENCES

American Bankers Association. 1984. Statistical Information on the Financial Services Industry. 3rd Edition. Washington, D.C.

Business and Health. 1986a. Insurer provider networks: A marketplace response. (January-February):20-22.

Business and Health. 1986b. The world of insurance: What will the future bring? (January-February):5-9.

Collier, D. 1983. The services sector revolution: The automation of services. Long Range Planning 16(December):10-20.

Collier, D. 1984. Managing a service firm: A different game. National Productivity Review (Winter).

Cooper, K., and D. Fraser. 1984. The changing structure of the financial services industry. In the Banking Deregulation and the New Competition in Financial Services. Cambridge, Mass.: Ballinger Publishing Company.

Credit Suisse-First Boston Bank. 1983. Section 1.3 in Euromoney Yearbook 1983. Boston, Mass.

Deardorff, A., and R. Jones. 1985. Comparative advantage and international trade and investment in services. Fishman-Davidson Center discussion paper. Philadelphia, Pa.: The Wharton School, The University of Pennsylvania.

DeYoung, G. 1985. Health care looks beyond the hospital. High Technology (September).

The Economist. July 6, 1985. The other dimension: Technology and the City of London. A survey. 296(7401):50.

The Economist. March 16, 1985. International investment banking: The world is their oyster. A survey. 294(7385):58.

The Economist. June 29, 1985. A threatening telephone call from the computer company. 295(7400):69.

The Economist. November 29, 1985. Telecommunications: The world on the line. A survey. 297(7421):62.

Financial Times. October 4, 1985. Into the era of specialization, special Financial Times survey on computing services.

Financial Times. October 4, 1985. Keeping ahead through sophistication.

Fishman-Davidson Center. 1985. The Service Bulletin (summer). Philadelphia, Pa.: The Wharton School, The University of Pennsylvania.

Gardiner, R. 1985. Sears' role in consumer banking. The Bankers Magazine (January-February):6-10.

Grimm, W. T. 1984. Mergerstat Reviews. Chicago, Ill.: W. T. Grimm Co.

Grossman, G., and C. Shapiro. 1985. Normative issues raised by international trade in technology services. Fishman-Davidson Center discussion paper. Philadelphia, Pa.: The Wharton School, The University of Pennsylvania.

Kirkland, R. 1985. Are service jobs good jobs? Fortune (June 10).

Koch, D., and D. Steinhauser. 1982. Challenges for retail banking in the 1980's. Economic Review (May). Federal Reserve Bank of Atlanta.

Kutcher, B., and J. Mark. 1983. The services-producing sector: Some common misperceptions reviewed. Monthly Labor Review (April):21-24.

Mark, J. 1982. Measuring productivity in the services sector. Monthly Labor Review (June):3.

Maslow, A. H. 1954. Chapters 5 and 8 in Motivation and Personality. New York: Harper and Brothers.

National Center for Health Statistics. 1980. U.S. Vital Statistics. Washington, D.C.

National Science Foundation. 1975. The consequences of electronic funds transfer: A technology assessment. (RANN) NSF/RA/X-75-015(June).

New York Times. October 27, 1985. Measuring the services economy. F4.

Office of Technology Assessment. September 1984. Effects of information technology on financial services systems. OTA-CIT-202. Washington, D.C.

Office of the U.S. Trade Representative. December 1983. U.S. National Study on Trade in Services. Washington, D.C.

Quinn, J. B. 1983. Overview of current status of U.S. manufacturing. In U.S. Leadership in Manufacturing. Washington, D.C.: National Academy of Engineering.

Quinn, J. B. 1986. Technology adoption: The services industries. In The Positive-Sum Strategy, R. Landau and N. Rosenberg, eds. Washington, D.C.: National Academy Press.

Roach, S. 1985. The information economy comes of age. Information Management Review (Summer):9-18.

Rule, J. B. 1975. Value choices in electronic funds transfer policy. GPO 041-001-00110-7. Washington, D.C.: Office of Telecommunications Policy, Executive Office of the President.

Sellers, W. D. 1985. Technology and the future of the financial services industry. Technology in Society 17:1-9.

Standard & Poor's Industry Survey. December 13, 1984. Banking and other financial services. 152(50):B18.

UNIPUB. 1984. Measuring productivity: Trends and comparisons. From the First International Productivity Symposium, Tokyo, Japan, 1983. New York: UNIPUB.

United States Banker. 1985. (August):10.

U.S. Department of Commerce. 1985. 1985 U.S. Industrial Outlook. Washington, D.C.

Coping with Technological Change: U.S. Problems and Prospects

RAYMOND VERNON

The consequences of technological change on economic activity are extraordinarily diverse, and any effort to single out a "dominant" consequence that will overwhelm all the others is certain to fail. But to think seriously about public policy responses requires making guesses about the nature of future problems, distinguishing the exceptional from the commonplace, and differentiating the dominant from the trivial. This chapter will begin, therefore, by examining a set of broad generalizations about the economic consequences of technological change over the last four or five decades, generalizations that reach beyond the available facts in some respects to create a basis for policymaking.

BASIC SHIFTS

One development that deeply concerns policymakers is an apparent secular decline in the relative world position of U.S. industry, a decline that is usually attributed to a fall in U.S. industrial competitiveness in world markets. The idea of a decline in the "competitiveness" of any given sector in a national economy is not easy to define and document. It is not yet clear whether or not we understand the nature of the structural changes that have been going on in the U.S. economy during the past 20 years or so.[1]

Consider, for instance, the changing role of manufacturing in the U.S. economy. Some data point to the conclusion that the manufacturing sector of the U.S. economy has declined in relation to the U.S. economy as a whole. For example, some shrinkage in the relative share of the U.S. labor force devoted to manufacturing occurred during the late 1970s and in the 1980s.[2] Over the longer term, the contribution of the manufacturing sector to the country's GNP, when measured in current dollars, fell from 28 percent in 1960 to 21 percent in 1983. But during the same period, the manufacturing sector's contribution, when measured in constant dollars, hovered around 23 percent without a clear trend (*Economic Report of the President*, 1985, p. 244). Data that purport to measure national output in "constant prices" over a period as long as two decades are inescapably vulnerable. Yet, those data suggest that the apparent relative decline in manufacturing was due to a relative fall in the prices of manufactures rather than to a fall in real output. If that suggestion were based on more robust figures, it would put a wholly different complexion on the seeming decline in U.S. manufacturing over the past two decades.

Those who believe that U.S. industry is losing in competitiveness also point to a decline in the importance of U.S. merchandise exports relative to the exports of all developed economies, a trend that has included such important sectors as chemicals, machinery, and automobiles (Lipsey, 1984, p. 70; United Nations, 1983, pp. 1046, 1048). Are these developments consistent with the idea of a decline in the competitiveness of U.S. manufactures?

Since 1945 the number of producers and consumers on the world scene has undergone an extraordinary increase. Outstanding among this new contingent of producers and consumers have been the Japanese people and the populations of the various newly industrialized countries—notably, Korea, Taiwan, Hong Kong, Mexico, Brazil, and even India. And soon China will be joining the list.

The emergence of these new producers and consumers in world markets has depended on various factors, including some extraordinary improvements in the technology of transportation and communication—improvements that have greatly reduced the costs to these late-industrializing countries of assimilating information from foreign countries and of moving goods and people across great distances. Once the information barrier was overcome, the comparative advantage of these countries shifted in favor of manufacturing activities, a shift that in the first instance was based largely on their wage structure in industry.

The same technological changes in communication and transportation have contributed to the spread of U.S.-based multinational enterprises over the past four decades. Technological advances have reduced the cost of search for new locations on the part of multinational enterprises. And as multinational enterprises have gained experience at such locations, their responses to subsequent opportunities have been quicker and more assured.

To be sure, some multinational enterprises have drawn back from various foreign locations in recent years, reacting to political threats or poor profits or both.[3] But, over the long run, U.S.-based enterprises have been drawing on foreign facilities with increasing alacrity as their experience with such facilities has broadened and deepened. For instance, a systematic study of the behavior of 57 manufacturing enterprises in the period from 1945 to 1975 dramatically illustrates the ratcheting effect of past experience on the disposition of those enterprises to set up additional facilities in foreign countries. In the decade after World War II, U.S.-based multinational enterprises that were entering on the manufacture of a new product typically allowed a considerable number of years to pass before setting up a manufacturing facility for the new product in a foreign location. By the 1970s, however, that lag was typically much shorter; and the greater the foreign experience of the firm, the more dramatic the shortening of the lag (Vernon and Davidson, 1979, pp. 37-62).

The reactions of multinational enterprises to the revolutionary improvements in communication in recent decades have taken other forms as well. Such improvements have allowed enterprises to manage their various subsidiaries as related units in a unified system rather than as isolated free-standing investments. And the unitary approach has encouraged some multinational enterprises to plan their financing, production, and marketing activities on a global scale, thereby increasing the complementarities and interconnections among the various affiliates that made up the enterprise (Casson and Norman, 1983, p. 63; Porter, 1985, pp. 410-414; Teece, 1983).

The impact of technology on the location of economic activity, however, has not been confined to multinational enterprises. With a relative decline in the costs of communication, the costs of transferring technology between independent parties can be presumed to have gone down, whereas the speed of such transfer has probably increased substantially. That development has been particularly important because of a concurrent development during these same decades—namely, an increase in the number of sources from which the technology required in any given product line could be drawn. Various studies demonstrate that as industries have matured, the number of firms engaged in the industry has tended to grow, as has the number of suppliers of equipment to such industries.[4] As a consequence, newcomers aspiring to enter the industry are in a position to shop around among a larger number of firms that might be able to provide the necessary technology.

The increase in the number of firms engaged in these technological transfers, whether as donors or recipients, has also been due in part to the efforts of various governments to support national champions—firms that could represent their countries as entrants in some given product line. In the industrialized countries of Europe and in Japan, these national champions have typically been found in the technologically advanced sectors, such as biotechnology industries, computers, or commercial aircraft. In the newly industrializing countries, such as Brazil or Korea, governmental support has been thrown to the more mature industries, such as chemicals and metal fabricating.

To launch this new generation of national champions has not always been easy; when governments have been successful, their success has usually been attributable to various policies, including a willingness to invest heavily in education and in the assimilation of foreign technologies. In many cases, the ability of the national champions to compete in technologically sophisticated lines of manufacture has also been helped by the import protection their governments provided in their home markets, as well as by the selective provision of various means of support.[5]

Many observers in the United States attribute the relative decline of U.S. industry during the past few decades to the unwillingness or inability of the United States to engage in policies of a similar kind (Diebold, 1983, pp. 644-654; Lodge, 1984, pp. 3-31; Magaziner and Reich, 1982, p. 326; Thurow, 1980, pp. 191-214). In reply, foreign commentators have contended that the R&D programs of the United States have been so large as to dwarf the support provided by other countries.[6] Others have noted that the U.S. government has made extensive use of import restrictions during the past decade or so, matching or exceeding the import restrictions of the Europeans and Japan.[7]

It may be that the largest part of the new competition faced in recent decades by the United States was almost inescapable. A necessary condition contributing to that development was the technological improvements in communication and transportation, a force that dramatically reduced the costs to foreigners of acquiring technical skills and technological information. With these new opportunities universally available, some countries, such as Japan, Korea, and Brazil, exploited the new situation more effectively than others. But a substantial global rise in competition, especially in the more mature product lines, could not have been avoided.

To be sure, the entry of a wave of new industrial producers—many of them protected in their home markets and bolstered by subsidy—created significant problems. In some industries, the more efficient production facilities of the newcomers made existing facilities in the United States redundant; this was a major factor in Japan's early reentry into the world steel market, for instance, as well as the later entry of the newly industrializing countries not only in the steel market but also in many lighter product lines. Because many of the industries that were involved were highly capital-intensive and characterized by substantial economies of scale, the appearance of the new entrants commonly produced a state of overcapacity, as in the case of steel, basic chemicals, and petroleum refining.

Some observers have attributed the relative decline in the share of U.S. manufacturing industries to other factors as well. None of these, in my opinion, is as important as the decline in the costs of acquiring technology and the concurrent starting up of industrialization programs in so many countries. One common contention, for instance, is that the capabilities and attitudes of U.S. managers have changed in recent years, reducing the capacity of U.S. firms to produce competitive products. Managerial attitudes and other cultural factors are no doubt relevant in a diagnosis of national business behavior; obviously, the values of British or German or Nigerian businessmen have something to do with their performance. But one should recall that throughout most of the 1950s and 1960s, observers were typically extolling the special virtues of American entrepreneurship, including its attention to profit and its swift responses to changes in the environment.[8] Ten years later, some of these same characteristics were being condemned, charged with bringing about a decline in U.S. business (Abernathy and Hayes, 1980; Thurow, 1980, pp. 3-25). This rapid shift in perception raises questions about its credibility.

Moreover, the global position of U.S.-based multinational enterprises, including their foreign affiliates, offers a different impression of U.S. managerial performance than does the output data for the United States alone; if the global exports of U.S.-based enterprises are taken as a guide, no competitive slippage occurred between 1966 and 1977 (Lipsey and Kravis, 1985). The apparent ability of these multinational networks to cling to their share of world export markets casts added doubt on the hypothesis that a deterioration in the quality of management explains the shrinkage in the U.S. position.

A more serious contention with respect to the role of U.S. management in the asserted decline of U.S. industry is the view that U.S. management practices—epitomized by the Ford assembly line and time-and-motion studies of the 1920s—have become obsolete, outdistanced by the more flexible management techniques of other countries, notably Japan (Reich, 1983, chapter 4). Undoubtedly, certain factors in the Japanese environment have contributed to strikingly high levels of productivity in some sectors of that economy (Brooks, 1985; Roy, 1982). The problem, however, is one of diagnosis. What is the critical variable? Is it the high levels of literacy of Japanese workers; the lifetime employment practices of the large firms; the recruitment and promotion practices relating to managers; the extensive use of subcontractors to provide flexibility; the financial linkages of the big firms to private banks and public credit intermediaries; or some combination of the above? The basis for assessing the relative importance of these various factors is very slim. Besides, the competitive pressures coming from countries such as Korea or Brazil, where managerial practices differ from those of Japan, also have to be explained. Accordingly, the part played by managerial practices on the relative decline of U.S. industry continues to be uncertain.

Another explanation that some analysts favor for the relative decline in U.S. industry is the effect of an overvalued dollar. This condition has plagued U.S. exporters since the early 1980s and has contributed to a significant increase in U.S. imports.[9] But the allegations of a U.S. decline in competitiveness long antedate the recent period of marked overvaluation of the dollar; the long-term fall in the U.S. share of world merchandise exports, moving from 16.5 percent in 1955 to 12.2 percent in 1975, was already being observed and commented upon in the 1960s (Lipsey, 1984, p. 70). No doubt the overvaluation of the dollar in recent years has added to the difficulties of U.S. manufacturers. But it would be prudent for policymakers to assume that the overvaluation simply aggravated a tendency that had deeper causes and longer antecedents.

In considering what policy responses might be desirable for the future, one cannot escape the need for projections. For instance, it seems plausible to assume that the rapid rate of industrial change will continue, requiring a continuous change in product lines. To be sure, some factors may slow the rate a little. For one thing, the initial surge of investment in Japan and the developing countries after World War II was fueled by policies of import substitution; and that factor is likely to play a lesser role in the future as the limits of import substitution are approached. In addition, Japan's entry into new product lines now requires her producers to take a much greater share of the risks and burdens of innovation, rather than depending primarily on absorbing and improving on existing technologies. The newly industrializing countries also face new financial constraints that may slow the rate of change; international lenders have grown much more cautious and governments much more chastened in planning for additions to industrial capacity.

Nonetheless, the U.S. economy must count on continued rapid industrial change. Despite the widespread current sentiment in the United States for protection against imports, the probability that U.S. government policies can effectively slow such change appears to be fairly low. With multinational enterprises in the United States accounting for two-thirds or more of U.S. industrial output, most U.S. producers will be ceaselessly looking for opportunities to reduce their costs; and as the history of electronics, automobiles, and steel graphically suggest, an important aspect of the

adjustments to competition that such multinational enterprises will undertake is to create offshore sources for some part of their output.

Are U.S. import restrictions likely to arrest this trend? Any U.S. decision
to restrain such activities would have to overcome the opposition of a
considerable sector of U.S. industry and could be implemented only with
draconian measures of restriction. It is inconceivable, for instance, that
U.S. policy would include prohibitions on U.S. firms from creating subsidiaries abroad to serve their markets in third countries. (Indeed, under
present circumstances, it is hard to conceive of any market economy
effectively imposing such a prohibition on its enterprises for very long,
unless the enterprises are owned by the state.) Assuming that the conditions
of rapid change continue, then, what problems are foreseen, and what
policies seem indicated?

THE PROBLEM REDEFINED

Many economists have been skeptical that the apparent decline in the
relative position of U.S. manufacturing poses any serious problem in public
policy. They are quick to point out that, when a country loses markets to
foreign competitors, those very losses initiate a series of internal and
external adjustments that eventually make up the losses at levels of output
that maximize national welfare. This demonstration, however, typically
sets off a reaction of the deepest frustration on the part of politicians,
newspapermen, and other ordinary mortals, and a reaction of the deepest
annoyance on the part of other members of the economics fraternity,
because of various critical and unanswered questions:

- Are the economic costs associated with the consequent adjustments
  being distributed to various groups in the economy on a pattern that
  is politically tolerable?
- Is the new production mix conducive to the continued future growth
  of the economy?
- Is the new production mix adequate for national defense requirements?

The Distribution of Costs

The problems and opportunities created by the increasing openness of
the U.S. economy have fallen unevenly among the various U.S. industries.
In other countries, such a disparate distribution might not have been so
disturbing. In Japan, for instance, the remnants of the old zaibatsu organizations have created interest groups or conglomerates who characteristically embrace both rising and declining industries. Besides, operating
according to a pervasive sense of "fair shares," Japanese policymakers

have tended to limit the size of the adjustment burden that is placed on any group or industry (Vogel, 1979, pp. 117-124). In Europe, an official readiness to provide ad hoc support to faltering industries and laggard regions creates the appearance—and perhaps even the reality—of lesser risk for industrialists and their workers.

Among U.S. industries, some of those beleaguered by foreign competition have had the prescience and the resources sufficient to escape from their U.S. locations and to set up producing facilities in lower-cost locations overseas; but others have failed to take such steps. According to various analyses, certain U.S. industries have been especially vulnerable to foreign competition because they have been paying wages in the United States well above the levels that seemed indicated on the basis of their productivity relative to other U.S. industries; in that category notably were automobiles and steel.[10] A second group of affected industries were those that had been insulated from international competition by the frictional costs of distance, costs that had previously restrained producers from acquiring the necessary technical capabilities and from learning about foreign market opportunities; illustrative of such products were various consumer soft goods and electronic items.

Some U.S. industries, however, have benefited from the shrinkage in the frictional costs of distance. One such industry has been the mass merchandisers of consumer products. With great rapidity, distributors such as Sears used the changed situation to search out low-cost foreign sources of consumer products, such as small black-and-white TV sets, for importation into the United States. In addition to the mass merchandisers, the U.S. industries that have derived advantages from the increased openness of the U.S. economy have been notably the exporters of capital goods, intermediate materials, services, and management skills. There are numerous indications that these offsets have been very large, indications reflected in the expanding importance of the export of technical services from the United States and of the flow of interest, dividends, and royalty income to the United States.[11] But the U.S. industries that have profited from those expanded sales have not been the same as those affected by the increased competition from abroad.

At least as important as the distribution of costs and benefits among U.S. industries, however, has been the distribution of costs and benefits among classes in the United States. Some U.S. residents see themselves as helped by the trend toward open markets because they can offer their services or their capital globally, thereby broadening their opportunities and their income. Other U.S. residents see themselves as hurt by the trend, in relative if not in absolute terms, because their jobs are displaced by increased foreign exports.

To be sure, some of the displacements forced on U.S. residents may have contributed so much to U.S. welfare as a whole as to justify the inequality in the burdens of adjustment. For instance, workers in industries in which relative wage levels have not reflected relative productivity, such as those in automobiles and steel, have in effect been extracting a rent from the rest of the U.S. economy. The reduction of that rent can be seen as a welcome contribution to the country's aggregate welfare, including the welfare of low-income groups. Moreover, increased international competition has probably helped low-income groups in the United States disproportionately in the effects of such competition on the prices of clothing, household appliances, food, and other staples. Yet the polarization issue still remains: The increasing openness of U.S. markets appears to reduce the going wage for factory labor as well as the ability of their unions to bargain. Meanwhile, the growth of the U.S. position in the export of capital, services, management, and high-tech products provides the most obvious income benefits for relatively high-income groups.

The Requirements for Future Growth

As suggested earlier, economists are likely to take somewhat different approaches from politicians in identifying the future problems of the United States. Economists tend to look on the U.S. economy as endowed with a given bundle of resources and to ask whether the U.S. economy, given its potential, is maximizing its opportunities for growth. Politicians, on the other hand, usually make no explicit assumptions about the U.S. potential and tend to ask if the U.S. economy is keeping up with the growth of other countries, notably Japan. In political terms, it does little good to compare the U.S. economy's performance with some full-employment potential or to observe that its slippage in position for the most part involves relative reductions in income, not absolute declines. In policymakers' eyes, the objective is not only one of equilibrium but also of a growth rate that matches the rates of other industrialized countries. U.S. policies in the field of technology are bound to reflect that preoccupation.

With the prospects for future growth in mind, policymakers frequently express the concern that the U.S. economy will be turned into a "service economy," a phrase that conjures up a picture of an endless succession of ski resorts and hamburger stands. As noted earlier, the manufacturing component of the U.S. economy has been shrinking by some measures, holding its own by others. In any case, it is not at all clear that a shift toward a service economy would prejudice U.S. prospects for growth. A great deal depends on the nature of the services. Some services allow an economy to capture large rents, such as brain surgery and Disney World's

EPCOT Center. And other services, such as research in biotechnology, are building blocks for future growth. Nonetheless, some analysts argue that the service industries, especially when separated geographically from manufacturing activities, are in general a poor vehicle for capturing rents or ensuring future growth.[12]

Questions of this sort take us into uncharted waters, in which none of the diverse viewpoints is likely to represent more than an informed guess. Yet policymakers find themselves obliged to shape their policies on the best evidence available, even if the evidence is flimsy.

Requirements for Defense

Nations are doomed to prepare for the wrong war. No nation was much prepared for the trench warfare of World War I, the war of movement of World War II, or the jungle warfare of the Vietnam era; and none is likely to be much prepared for the next. One factor in preparedness is the altogether understandable desire for a minimax strategy. Of the various hypothetical possibilities, the assumption that the next war will be much like the last, however implausible that may be, still ranks very high. Hence the idea that a country will need, for example, a national steel industry, a national aluminum industry, and a national petroleum industry for defense purposes is bound to be an unshakeable element in national defense planning.

If there were any significant probability of a prolonged conventional war on a large scale, the case for maintaining a large core of such industries might be reasonably strong. But conventional warfare on the Korean or Vietnamese scale, extensive as these were, poses almost no problem for the U.S. mobilization base, past or prospective. And if nuclear warfare is contemplated, the maintenance of a core of industries is almost irrelevant to the actual conduct of the war. On security grounds, therefore, the case for maintaining the usual core of basic industries is not obvious.

The case for maintaining a large core of "defense" industries would be particularly weak if the policy were to interfere with another security objective of obvious importance, namely, that of maintaining a flexible work force, capable of technical improvisation and adaptation on a large scale. That need was evident even in the conventional warfare of World War II, when the German and Japanese economies managed to support a large-scale war effort for a sustained period despite the extensive damage done to their relatively slim industrial base (Schultze, 1973, p. 526; Vernon, 1955, pp. 77-88). In a world in which nuclear warfare cannot be excluded as a possibility, the capacity for adaptation and flexibility should be a paramount security objective.

Here again, more research is needed. But one must be realistic about the extent to which any research on this point can influence policy. Whatever the outcome of such research, it seems inevitable that the political process of the United States and of other countries will require policymakers to cover altogether implausible contingencies, such as the possibility that World War II will be repeated.

The Limits of Policy

Upon identifying the main issues, one is propelled to a basic conclusion: Given the nature of the issues, better data and closer analysis are unlikely to have more than a marginal effect on the behavior of the U.S. government and other governments. The problems are too large and too conjectural, and the domestic and international mechanisms for action too feeble to generate more than marginal impact. Yet as one sifts through the issues raised in this summary, there are three areas in which some marginal influence is possible and desirable.

One such issue is how the United States can enrich and enlarge factors of production that are likely to operate on U.S. soil, especially the quality of its labor and management. A second issue, intimately related to the first, is how a tendency (or the appearance of a tendency) toward economic polarization within the U.S. economy—an apparent consequence of the economy's increased openness—can be arrested and reversed. A third issue, inseparable from the first two, is how those objectives can be achieved without beggaring other countries and thus setting in train a process that in the end would bring down the U.S. economy as well.

ENHANCING THE FACTORS OF PRODUCTION

The Educational Challenge

As nearly as any analyst can tell, the ability of the U.S. economy to maintain a high living standard relative to other countries will depend on its being able to develop a literate and flexible labor force: literate so that it can perform complex and demanding tasks; flexible so that it can take on new tasks as products and processes continue to change. There is nothing new in that generalization. Historians usually explain the spectacular rise of Germany and Japan as well as the relative decline of the United Kingdom in part by reference to their respective systems of education (Landes, 1969, pp. 340-347). And if that factor was important in the environments of 50 or 100 years ago, it is many times more important today.

Historically, the United States has provided its nationals with extensive facilities for basic education. And it has led the way among the advanced industrialized countries in providing opportunities for higher education, at least as measured by the number and proportion of persons attending institutions of higher learning. Although illiteracy has been endemic in southern rural areas and city slums, that fact has not been sufficient to prevent the country from maintaining its high living standards.

Increasingly, however, the quality of education in the United States is a matter of national concern. Part of that concern is with the absolute quality of the education provided at all levels, given the demands of modern society, and part is with the relative position of the United States as other countries extend and improve their education systems.

In responding to that concern, the country's federal structure and traditions of local rule demand that its states and localities exercise a considerable role in determining the quality and content of education programs; but the consequences of their decisions are bound to be national rather than merely local. If Kentucky substitutes basketball for geometry, the consequences are felt in California; and when the central cities fall behind the suburbs in providing educational facilities, the suburbs cannot escape some of the consequences.

Accordingly, the dramatic reduction in the federal government's oversight role in education during the past half-decade, although a logical corollary of the New Federalism concept, is also deeply worrying in its implications. To be sure, some localities have responded with sharply improved educational efforts. But the variations in educational services from state to state and between city and suburb are expected to grow, not decline. And the continuous influx of new immigrants could exacerbate these distinctions by placing heavy burdens on selected cities and states such as Los Angeles and New York.

The problem is not solved by simply reestablishing some measure of federal oversight in the field of education. In the decade of the 1970s, while federal oversight was being strengthened and extended and while federal resources were being provided to supplement local financing, school curricula nevertheless deteriorated, and illiteracy in the United States probably increased. Yet as long as primary and secondary education is financed and controlled largely at the local level, there is almost no hope of bringing large pockets of the population up to the standards of literacy and flexibility that are consistent with the maintenance of the world's highest living standards.

The problem is not limited to the fact that the U.S. labor force may be unable to provide the productivity that is consistent with high living standards for the country. A condition of that sort is bound to sharpen the differences between those who draw their income from operating in a world environment and those who rely on the U.S. environment for their income. Polarization in income and polarization in political outlook are likely to go hand in hand, leading to increasing bitterness over issues of transfer payments within the country and issues of international economic policy at the borders (Verba and Orren, 1985, pp. 174-177). Of the various challenges that the United States faces for maintaining a lead in the modern world, this one heads the list.

Improving Industrial Innovation

Practically every econometric study of the sources of growth of the U.S. economy ascribes a considerable part of that growth to some amorphous factor beyond capital or labor, usually thought to be an amalgam of improved technology, improved organization, and increased education (Denison, 1979, pp. 104-107). Not surprisingly, therefore, the policy of the United States has been to encourage research and development activities on the part of industry, using tax credits and military procurement programs as the principal vehicles for official support.

That government policies are capable of stimulating successful research and development is clear beyond question. The extraordinary increases in agricultural productivity in the United States and elsewhere during the past 50 years, although reflecting the joint efforts of the public and the private sectors, would have been impossible without the public contribution (Nelson, 1982, pp. 251-252). The U.S. aircraft industry has been overwhelmingly dependent on the military budget for much of its research and development; although various factors help to explain the performance of the civilian aircraft producers, including the existence of a large internal market, the support of the U.S. defense establishment has been a critical part of the mix. The technology of extraterrestrial space represents another partnership in which the public contribution has been critical.

Not all public initiatives have been successful, however. Some have foundered on technical grounds, as in the case of the "transbus," and others because of a misreading of the market, as in the case of synfuels. Moreover, public initiatives such as military procurement sometimes generate costs that inhibit commercial research and development rather than support it. The principal means by which military programs encourage commercial research and development is through the spillover effects to nonmilitary applications. But military hardware is becoming highly specialized, being developed in systems that are increasingly remote from civilian applications; whereas the bombers of the 1940s and 1950s could be adapted eventually to commercial aircraft needs, the ICBMs of the 1980s offer no analogous possibilities (Brooks, 1986, pp. 134-135). In addition, some analysts are of the view that U.S. military programs on balance raise the cost and reduce the availability of engineers and other technicians for commercial enterprises in the United States (Brooks, 1985, p. 349). Since the U.S. pool of engineers and technicians is smaller than that of Japan or Germany in relation to the total labor force, the United States is obliged to divert a substantial part of that pool to innovations that characteristically have no direct civilian use.

A larger problem associated with the stimulation of research and development on the part of U.S. enterprise, however, applies not only to the stimulation provided by military programs but also to the stimulation generated by tax credits. The propensity of technology to cross national boundaries has been growing rapidly, mainly as a result of the improvements in communication and transport. That movement takes place through numerous channels: for instance, through word of mouth, patent applications, scientific meetings, licensing agreements, and communications among the affiliates in multinational networks. Accordingly, R&D programs whose costs are borne in part by the U.S. budget in the hope of increasing U.S. productivity eventually serve to increase the productivity of other countries as well, including that of competitors. To be sure, those benefits are not altogether lost on the U.S. economy, given the intimate relationship between the prosperity of other countries and that of the United States. But it cannot be assumed that the benefits of that increased productivity are fully returned to U.S. investors; more likely, they are shared with foreign workers, foreign governments, and foreign consumers. In any case, those whose eyes are fixed on U.S. adjustment problems resulting from the openness of the U.S. economy should not suppose that the stimulation of research and development by U.S. firms necessarily reduces those problems.

The implications of the fact that technology cannot easily be locked onto any single national turf have not yet been fully assimilated in the minds of science policymakers anywhere in the world. Once policymakers in the United States grasp the point, they may be tempted to conclude that the U.S. government should withdraw its support from efforts to stimulate research and development. Yet if such a policy had been adopted as a guiding principle 50 years ago, the world's food production would be much lower today, and the extraterrestrial communication satellite would still be a remote vision. If the fruits of our technological efforts must inevitably be shared, however, there is a case for encouraging joint national programs in the stimulation of technology rather than unilateral national efforts, wherever that can be arranged. True, joint national efforts usually generate formidable organizational problems that national efforts can sometimes avoid; but confronting those problems may be seen as preferable to assuming the full cost of research activities whose benefits, in the end, will be shared with other countries.

In any case, the United States may have something to learn from the efforts of other countries to stimulate research and development. Numerous reasons have been cited why the efforts of other countries cannot always serve as a useful guide for policy in the United States. One reason commonly adduced for avoiding such policies is specious, but a second has to be taken more seriously.

The specious reason is that if it were economically wise for enterprises to risk their funds in developing a given innovation, one or more enterprises would already be attempting to develop it. As numerous analysts have pointed out, there is a considerable risk that the private sector will underinvest in research and development, especially when the private firms are unsure whether they will be able to capture the profits that the invention might generate (Bozeman and Link, 1985, p. 376; Freeman, 1982, p. 168; Mansfield, 1968, p. 187). This is an acute problem, for instance, in products or processes that are not easily protected by patents, or in products that might prove to have a very short life, such as memory chips. In such cases, it may pay for the government to reduce some of the perceived risk by guaranteeing a minimum return in effect to the successful inventor.

The second ground on which many U.S. policymakers refuse to entertain proposals for the public support of selected development targets is that the U.S. government is inherently unable "to pick winners and losers." Sometimes the contention is made as if it were axiomatic, an inescapable consequence of the characteristics of the system; but it is unclear why a system that is capable of producing the moon landing, agricultural research, and the space shuttle is inherently incapable of successfully identifying research goals that are technically feasible and socially profitable. Once one accepts that the market may be failing to provide reliable signals for profitable R&D investment, there is no reason to exclude the possibility that the government might do better.

Stimulating Management

With the swift changes of the past few decades in the structure of markets and enterprises, the U.S. landscape is littered with anachronistic regulations and outmoded precedents that tend to put a damper on managerial initiatives. Among the most obvious are those relating to antitrust and restrictive business practices.

U.S. antitrust policy is built in part on the principle that maintaining competition depends on avoiding high levels of concentration of U.S. production among a few firms in any given product line.[13] But as markets have become global in scope, and as multinational enterprises have increased in number, the standards by which U.S. law and regulation have traditionally judged the effects of mergers and consolidations on competitive conditions in the United States have grown less relevant. Obviously, imports have to be taken into account in such an evaluation, a fact that lawmakers are just beginning to appreciate. But imports alone do not begin to reflect the changed competitive conditions of world trade and investment. The degree to which Fiat's existence acts as a brake on General Motor's market behavior in the United States cannot be judged simply by the level of U.S. imports of Fiat cars (which is small). Instead, as executives in the United States design their product and pricing policies, they are now obliged to consider the possibility that Fiat (or Toyota or Saab or Peugeot or Hyundai) might use their vastly increased capability for movement by establishing a production plant in the United States. As the statistics of foreign direct investment in the United States dramatically illustrate, that possibility is now very real.[14]

This profound change in international markets suggests the need for a radical overhauling of the standards and precedents of which the U.S. antitrust authorities are captive. It suggests the possibility of a more relaxed policy toward various forms of cooperation and mergers among U.S. producers in cases in which either foreign imports or foreign direct investment seem likely.

It does not follow, however, that antitrust problems are declining in intensity in this new global environment. In fact, some of the new interfirm arrangements that are appearing in response to these changed conditions raise novel questions that could prove important in the decade ahead. During that decade, the world may well see a period of relatively slow growth. In that event, leading producers who are somewhat insulated from competition will be understandably reluctant to spend on the creation of new technologies or to apply all the technological advances that their laboratories are capable of producing.

In the 1930s that combination of circumstances generated a substantial crop of international cartels covering electrical, chemical, and mechanical products, many of them dedicated to holding back the introduction of new products and new processes (Hexner, 1946, pp. 203-349). Today it is unlikely that any arrangements among producers dedicated to such objectives would take the transparent and explicit forms that the cartels of the 1930s sometimes adopted. Since then, the antitrust programs of governments in the United States and Europe have increased measurably in potency and sophistication. But the recent proliferation of joint ventures among industry leaders of different nationalities raises the possibility that leading firms may once again attempt to control the introduction of new technologies.[15] Parties to such alliances, it is true, usually have an initial intention of pooling their technology in the interest of increasing productivity or generating new final products, not of holding back on technological advances. But one must realistically entertain the possibility that a spell of hard times will alter the original direction of the alliances, especially when the alliances have linked the technological leaders in any product line.

The risk that technological advances will be checked by cartel-like international agreements arises from another direction as well. Numerous agreements among governments limiting the international movement of various types of steel now cover so large a part of the world's steel output as to take on many of the characteristics of a world steel cartel. The slow growth in the world demand for steel, when coupled with the curbs on international competition, could easily have the effect of arresting the steel industry's investments in new steel products and processes. That possibility is increased still further by the movement of the U.S. steel industry toward diversification into other industries, a movement that is likely to reduce the interest and the resources of management devoted to improving the productivity of steel facilities.

The case of steel raises the question whether rapid technological change always must be counted benign, or whether in some circumstances it generates social costs that exceed its benefits.[16] Many who support the international steel agreements would argue in favor of such agreements in precisely such terms. But that is a judgment better made by persons whose interests are centered on the welfare of the national economy as a whole rather than on the steel industry in particular.

A critical question for the future is whether the steel situation represents a pattern that will be followed by other industries as the demand for their products levels off. In that case, the antitrust laws of the United States and other countries may prove irrelevant, even though the restrictive consequences of the arrangements may be indistinguishable from those generated by private cartels. Here again, we confront a problem for U.S. policymakers to which almost no attention has so far been given.

REDUCING ECONOMIC POLARIZATION

Efforts to reduce the tendency toward economic and political polarization that may accompany increased openness in the U.S. economy are desirable on both social and political grounds. Some of the measures that could improve the productivity of factors employed in the United States would at the same time contribute to reducing economic polarization. Broad-based support for elementary and high-school education, for instance, would contribute to both ends. But other measures also need to be considered.

One approach to this issue is to review and revise national policies that now influence managements to respond to foreign competition by creating new offshore facilities, rather than by attempting to increase their productivity and lower their costs in the United States. Of course, the possibility of increasing productivity in the United States in response to increased foreign competition is not always realistic; some old U.S. industries must give way to such competition if new U.S. industries are to flourish. But there have been enough successful cases of U.S. firms holding their ground on U.S. soil through productivity increases to suggest that more such possibilities exist; General Electric's comeback in the manufacture of diesel locomotives is a spectacular case in point.

At the margin, some policies in the United States may be tailored to encourage such a response. For example, the tax forgiveness policies of other governments in some cases mean that profits generated by the foreign subsidiaries of U.S. firms are taxed at levels that are lower than the profits generated in the United States, a fact that may tip the U.S. firm's response in favor of an overseas location. Cases of this sort will increase as corporate income tax rates begin to rise again in the United States. Because this is a complex and tangled subject, answers to this problem do not come easily; but the issue is worth further exploration. More generally, numerous U.S. policies and programs are worth reviewing to determine whether modifications might keep production on U.S. soil.

Any program that is aimed at reducing the polarizing effects of open international markets, however, will be inadequate if it fails to provide for some positive form of trade adjustment, that is, some substantial lubrication of the process by which productive facilities are redeployed in the U.S. economy. By now, the idea of trade adjustment has become a little shopworn. First adopted as policy in the Trade Expansion Act of 1962, the trade adjustment program failed to contribute very much to the adjustment process by the time it was abandoned in the early 1980s (Aho and Bayard, 1980; Richardson, 1983). The fact is, however, that the concept was abandoned almost before it was really tried. During most of the two decades in which the program was on the books, its contribution to the adjustment process was limited to extending the period in which laid-off workers received unemployment compensation. In the latter 1970s some more imaginative approaches to adjustment at last began to be undertaken; but within a few years of the commencement of such experiments, the concept was abandoned. Whether effective retraining and relocation are possible on a substantial scale in the United States remains an unanswered question.

Meanwhile, it should be remembered that the threat of increasing polarization is as much a problem of attitudes as an issue of the actual distribution of income. In this context, symbols count. For this reason, if for no other, the salaries and fringe benefits that the top managers of large U.S. corporations appropriate for themselves may be playing a role that is larger and more significant than the economic cost. On the basis of informal inquiries of knowledgeable observers, my impression is that U.S. managers are paid two or three times the amounts received by their Japanese counterparts in companies of comparable size and substantially more than their European counterparts.[17]

Those U.S. compensation levels cannot easily be justified as the impersonal outcome of the play of market forces. Such levels are in effect fixed by the managers who receive them; and the managers in each enterprise determine those levels by making comparisons with other firms whose managers are engaged in the same process. To be sure, any managerial group that choses not to keep up with the others would probably find some of its members, including valued ones, abandoning the firm. Accordingly, any individual firm has little incentive to drop out of the race of matching increase with increase. The situation, however, promises to exacerbate the division between managers and workers. Whereas workers see themselves as living in an atmosphere of givebacks and two-tier pay scales in order to remain competitive, management continues on its course, relatively undisturbed, except for the risk of takeover bids. The polarizing effects of a continuation of such trends deserves the closest attention of policymakers.

AVOIDING A BEGGAR-THY-NEIGHBOR CYCLE

In the context of U.S. politics, however, the problems of managing technological change are likely to be framed as problems in international trade policy, and the instrumental responses are likely to be measures for restricting trade. Managing technological change, therefore, embraces all the familiar issues associated with trade policy.

The Shift in U.S. Position

According to various counts, something like 300 bills were introduced in committees of the 99th Congress to deal with international trade problems of one kind or another. Practically all called for some measures of restriction on U.S. imports. Some provided for consultation with foreign countries as part of the process leading to import restrictions, and others called for uninhibited unilateral action. The extraordinary emphasis on unilateral action is one manifestation of the gravity of the present situation. The United States seems prepared to court the obvious risk—one could almost say, the near certainty—that unilateral measures by the United States will generate unilateral countermeasures by others, contributing to a spiral that is hurtful to all countries.[18]

The U.S. willingness to expose itself to that risk reflects several different factors, including the ignorance of otherwise informed Americans about the extent to which the well-being of the U.S. economy has become related to the well-being of the world economy at large. Even a superficial review of the content of our school curricula and our mass media suggests why that ignorance exists. Courses in history and economics have almost disappeared from high school curricula, contributing to a pervasive indifference or illiteracy on subjects of international economic relations. To be sure, the media give occasional attention to some political or military spectacular involving foreign countries, such as an international hijacking, or a Gorbachev-Reagan encounter. But the coverage of international economic events is almost nonexistent in the mass media, whether in superficial or in serious form.

Ignorance and indifference, however, are not the only reasons for the spate of trade-restricting bills. Another factor may be the polarization tendency mentioned earlier, the perception of the growing differences between the interests of those Americans whose stake lies directly in the growth of the world economy, such as managers and stockholders of multinational enterprises, and the interests of those whose livelihoods are linked more directly to the U.S. turf, such as semiskilled textile and electronic workers. With such polarization, some groups in the United States will be tempted to disregard the possibility that U.S. trade restrictions might initiate a series of events that will do considerable damage to all countries in the world, including eventually the United States itself. In the eyes of disadvantaged groups, those consequences might appear no worse than a set of policies whose costs have to be borne by them.

Still a third strand in the current wave of demands for unilateral action is a mounting impatience among the American public and the Congress with the behavior of other governments in international trade.[19] It is true that most governments are more active than the United States in devising special measures to improve the competitive position of their own industries, a reflection of a different approach to the role of government than exists in the United States. But some of the impatience of the American public and the Congress is based on the assumption—the largely erroneous assumption—that other governments are more careless about their international commitments than the U.S. government, a view that explains repeated references to restoring a "level playing field" in international trade. To be sure, most governments are prepared to exploit every loophole created by the copious exceptions in the General Agreement on Tariffs and Trade (GATT) and are sometimes prepared to disregard the agreement altogether (Patterson, 1983). In this respect, however, the behavior of other governments is not readily distinguishable from that of the United States, which has frequently enacted legislation in flat violation of the GATT's provisions (Grey, 1983).

The gap between conventional U.S. opinion about the behavior of foreign governments and cold reality is strikingly illustrated by the case of Japan. With regard to the GATT's formal requirements, that country today is probably as much in compliance as the United States (Saxonhouse, 1983). The principal difficulties that the United States now encounters in U.S.-Japanese bilateral trade relationships are of a kind that the GATT's provisions do not—and perhaps cannot—address: the yen-dollar exchange rate, the tight vertical structure of domestic Japanese industry, the attitudes of Japanese consumers, and so on (Bergsten and Cline, 1985, pp. 21-105). To deal with issues of that kind, we must begin thinking of new forms of international regimes and international agreements far more ambitious than those represented by the GATT.

Practically all multilateral agreements are also bedevilled by the problem of the free rider. With more than 160 sovereign states of various sizes and interests engaged in international trade, countries that will hope to profit from open markets without opening their own are bound to appear in considerable number. That was true in 1948 when the GATT was created, and it remains true today. In 1948, however, the U.S. government believed that its interests were being served if some of these governments could be persuaded to give lip service to their adherence to a GATT regime, even if their role of free rider was legitimated in the process. Accordingly, the developing countries that joined the GATT were in effect exempted from any significant obligation under the GATT's rules while being entitled to the rights of a GATT member. Today, however, the United States no longer regards that lopsided situation as being in its interest, especially as it applies to such rapidly developing countries as Brazil and Korea.

From the U.S. viewpoint, the remaining alternative to unilateral restrictions on trade is the increased use of persuasion or threat: persuasion aimed at convincing a group of like-minded countries that it is in their common interest to devise new rules of the game that are responsive to the demands for a "level playing field," or threats aimed at increasing the inhibitions of the free riders. The use of persuasion will not be easy. For instance, it should not be supposed that the advanced industrialized countries are in a mood for joining the United States in measures for the liberalization of trade. In recent years, indeed, the European Economic Community has been pursuing its own program of trade-restricting measures, a program that more than matches the U.S. measures in its restrictiveness. The challenge is to turn this trend around, lest the remnants of the existing trade regime disappear.

Constructive and Destructive Agreements

For several years, in fact, the U.S. government has been moving toward a set of arrangements that have pointed away from the principles of non-discrimination and universality. Some of these arrangements have been constructive and benign, in the sense that they have probably improved the position of the United States without damaging the position of other countries. Others have been grossly destructive by these standards. Heading the list of destructive measures are the so-called voluntary export agreements (VEAs) or orderly marketing agreements (OMAs), which at present protect a considerable part of U.S. industry.

The essential element of these agreements is that the U.S. government coerces an exporting country into restraining its exports of a given product to the United States. The U.S. government's objectives in having the exporter impose the desired restrictions rather than imposing U.S. import restrictions are reasonably clear: The U.S. government escapes some of the onus of imposing a restriction on imports, especially if the conditions are such that an import restriction would be in violation of existing U.S. law or in violation of international trade agreements, notably the GATT. The U.S. government can "target" specific threatening exporters rather than all exporters of a given product, thereby discriminating in a way that would violate the GATT if incorporated in an import restriction; and the U.S. executive branch can buy off congressional opposition without taking on a legislative struggle.

This practice has its obvious drawbacks. It allows the U.S. government to pretend that the restriction is not of its doing, thereby reducing the government's ability to extract programs of adjustment from the U.S. industries that are protected. It encourages corruption and evasion in the trade practices of foreign exporters, who commonly transship their goods through other countries as a way of entering U.S. markets. And it makes a mockery of the rules on international trade agreements. Worse still, it leads to a gross distortion of such agreements as the U.S. government and others seek to legitimate their practices by obtaining nominal absolution in the GATT. The headlong movement of the U.S. executive branch toward the expanded use of the coercive bilateral approach, incorporated in those VEAs and OMAs, could eventually destroy what is left of the tattered rules of the GATT, before any successor regime can be devised. Finally, the VEAs and OMAs, to the extent that they are effective, have the unfortunate consequence of placing monopoly rents in the hands of the exporters, rents paid by the consumers in the United States. The "voluntary" restriction of Japanese car exporters from 1982 to 1984, for instance, was a bonanza for Japan's car exporters, increasing their margins of profit and their financial resources well beyond the levels they would otherwise have achieved and encouraging the accelerated upgrading of Japanese automobiles for sale in U.S. markets (Gomez-Ibanez et al., 1983).

Although the VEAs and the OMAs have been destructive to international trade, it should not be assumed that all trade agreements that are less than universal in scope are necessarily destructive. Economists have long recognized that in some circumstances, trade agreements that are discriminatory in their application may actually foster trade and may contribute to the welfare of nonparticipating countries as well as of those that participate (Lipsey, 1960). The European Economic Community and the European Free Trade Area, for instance, were sanctioned by GATT members on that assumption; and although the Community's policies are fundamentally at variance with the nondiscrimination principles of the GATT, one can make a case that the Community's operations contributed on balance both to international trade and to global welfare.

The U.S. government also has negotiated a series of less-than-universal agreements that presumptively have had benign rather than malign effects on trade, including notably a series of codes that were negotiated under the aegis of the GATT in 1979. More than a half dozen codes were initialed at the time, covering subjects such as the sale of commercial aircraft, practices in government procurement, and the use of certain types of subsidies that affect international trade.[20] These codes have been open-ended in the sense that any member of the GATT has been free to subscribe, accepting the obligations and procuring the rights provided in the code. But the codes are discriminatory in the sense that GATT members choosing not to subscribe are likely to be denied the benefits under the code.[21] Ultimately, the signatories have proved to be primarily the advanced industrialized countries.

Although the GATT codes have been feeble instruments in the first years of their existence, they represent one of the few approaches that offer fresh promise of the development of a benign trade policy. To be sure, they are discriminatory in their application; but any country can cure the discrimination by ratifying the code. Accordingly, such agreements do not turn away once and for all from the principle of an open world trading system, a fact that has made it possible for the GATT structure to tolerate and even to foster such agreements.

The disposition of the United States to pursue trade-liberalizing measures where it can, without being blocked any longer by the principle of universality, is evident from a number of its other initiatives in recent years, such as its bilateral discussions with Japan aimed at added liberalization of the conditions of government procurement, its proposal for a free-trade area with Israel, its interest in pursuing a selective free-trade arrangement with Canada, and so on (Bilzi, 1985).[22] Its eclecticism in trade matters now begins to approximate that of the European Economic Community, which has never been restrained by the principle of universality, as well as the approach of Japan, which also includes numerous departures from that principle.

The approach, to be sure, courts some acute dangers. The line between trade-creating and trade-destroying agreements can be exceedingly fuzzy. Agreements that are ostensibly available for signature by any country can be constructed in ways that shut many countries out. But it is difficult to envisage any universal agreements on the GATT pattern that offer much hope of arresting the trend to protectionism. Instead of addressing the relatively simple question of the level of a tariff rate or the existence of a requirement for an import license, future agreements must take up such complex and subtle questions as the international effect of a domestic subsidy, or the international impact of a regulation ostensibly applied for reasons of safety or environmental control or the support of a lagging region. These are subjects in which facts are hard to come by, consequences are complex, and domestic politics play a dominant role.

Such factors may explain in part why the existing GATT codes have proved so feeble. The machinery required for any degree of effectiveness in agreements that deal with complex nontariff trade barriers is more nearly approximated by the institutions of the European Community, with its elaborate provisions for fact-finding and adjudication. So far, the minds of U.S. policymakers are far from considering any such ambitious possibilities. Instead, their next efforts are being directed at extending the GATT to the liberalization of services and of foreign direct investment, programs that seem even less appropriate to the GATT structure than dealing with the new wave of trade restrictions. We have scarcely begun, therefore, to fashion international institutions that can cope with the new problems in the field of international trade.

U.S. POLITICS AND U.S. GOVERNANCE

Before any constructive action can be envisaged among governments for dealing with the problems of adjustment and accommodation to technological change, the U.S. government itself must be prepared to sponsor the requisite international action. The conventional wisdom of the moment, supported by the results of Gallup polls and the introduction of protectionist bills in the Congress, is that such sponsorship is almost out of the question. My own analysis, however, indicates that the U.S. position is far less determined than conventional wisdom would suggest.

It has been repeatedly observed that U.S. policy is based on the outcome of struggles among interest groups rather than on the rational conclusions of thoughtful referees. On that basis, one could easily conclude that the U.S. political process was as likely to produce a decision in favor of promoting open international markets as at any time in modern history. To be sure, a surge of imports has added to the acute difficulties of many domestic industries. But U.S. industry as a whole—including some of the leading firms in the industries threatened by exports—has never been more involved in world markets; the dominance of multinational enterprises in the U.S. economy has no historical parallel. The stake of the large U.S. banks in maintaining open markets is also vital, especially if they are to hope eventually to collect some of their large foreign loans. And the capacity of labor unions to support a protectionist policy, while still strong, is diluted by three factors: the decline in labor union enrollment; the growth in the relative importance of nonmanufacturing workers in the labor union movement;[23] and the decline in the level of unemployment. If sheer interest-group calculations were determining the direction of U.S. trade policy, therefore, one might reasonably expect a different atmosphere than the one that now prevails.

One key reason for the striking protectionist sentiment in Congress and the public press is that a coalition of protectionist interests in the United States has succeeded in a strategy that they have pursued persistently over the past 35 years. The strategy has been fundamentally to ensure that the problems associated with increased international trade are always addressed on a case-by-case basis rather than systemically. That result has been achieved by a succession of statutory measures that have given aggrieved domestic industries increasing powers to initiate individual proposals for trade restrictions. As long as such issues arise case by case, the atmosphere surrounding discussions of trade policy is bound to be protectionist, and the costs of imposing import restrictions on the economy at large—costs such as the impact on consumers and the risk of retaliation—are likely to play a secondary role in the decision-making process.

From time to time during the past 50 years, despite the strategy of the protectionist alliance, the U.S. government has moved dramatically in the direction of liberalizing its trade regime. In each instance, that decision has been taken in a context in which the decision makers were in a position to weigh not only the benefits but also the costs associated with higher import restrictions. Such moves occurred with the adoption of the Trade Agreements Act of 1934, the Trade Act of 1948, the Trade Expansion Act of 1962, the Trade Act of 1974, and the 1979 Trade Act. In each of these instances, costs and benefits of the trade regime were framed and presented as a single package.

Those same acts, however, carried the seeds of the eventual undermining of the trade policy they supported, as a series of technical amendments made it increasingly easy for special interests to raise their trade problems on a case-by-case basis. Until those procedures have been curbed, U.S. policies will be acutely vulnerable to the initiatives of such special interests.

Nevertheless, the bold moves by U.S. government toward trade liberalization—moves taken in each instance with the acquiescence of the Congress—underline a fundamental point with regard to congressional behavior. Conventional wisdom usually takes it for granted that the Congress will be protectionist, on the easy assumption that congressmen are doomed to bow to the interests that have been mobilized in their districts. But the record itself, as one or two studies indicate, is much more complex (Ahearn and Reifman, 1984; Pastor, 1980, pp. 186-199). Confronted with major choices that coupled the benefits with the costs, congressmen have been prepared to resist the pressures of specialized interests in their districts. In their handling of individual cases, administrators continue to be vulnerable to the tactical power and tactical skills of the interests. Accordingly, those who are skilled in broken field running through the sprawling governmental apparatus can often avoid the full exercise of the system of checks and balances to achieve their desired results in individual cases. The challenge is to find ways of restraining those possibilities so that the larger interests of the United States can play an appropriate role in the development of trade policy.

Because U.S. institutions and processes are so vulnerable to special interests in the handling of individual cases, some analysts have resisted the idea of having the U.S. government pursue selective policies of industry targeting such as has been seen in Japan, France, and Korea. The fear is that the U.S. process would produce results that were unrelated to any rational analysis of the individual cases. On the record, that is not an unreasonable concern. Yet, the swiftly changing character of the economy in which we live implies that with increasing frequency, individual enterprises may encounter transitional problems that threaten their very existence. In theory, such problems may be solved in various ways, including merger and acquisition. But the experience of the U.S. government in helping Lockheed and Chrysler bridge their transitional difficulties a decade ago suggests that a role for government is not to be excluded. Accordingly, the U.S. government may have to build new capabilities into its system of governance that allow it to deal rationally with critical individual cases as they emerge. Developing that capability is perhaps the most difficult of the various challenges discussed in this chapter.

ACKNOWLEDGMENTS

I am grateful for the critical comments provided by Harvey Brooks, Roger Porter, Robert Reich, and Dani Rodrik, and for the research support of Debora Spar.

NOTES

1. An excellent summary of the arguments on both sides of the issue is found in Norton, 1986. Underlying studies particularly worth consulting are *President's Commission on Industrial Competitiveness*, 1985, and Lawrence, 1984.
2. Between 1975 and 1983, the absolute size of the work force in U.S. manufacturing actually expanded by 15 million jobs, but its size relative to the total work force slipped to 18.0 percent from 20.9 percent (U.S. Bureau of the Census, 1985, p. 390).
3. From 1980 to 1984, foreign-direct investment as reported by U.S.-based firms has been virtually unchanged. See *Economic Report of the President, 1985*, p. 349, and *Survey of Current Business*, January 1986, p. 24. But that is probably the net result of a reduction in liquid assets and an increase in fixed assets abroad, as well as the revaluations of foreign assets related to an appreciating U.S. dollar.
4. A classic study of this phenomenon, covering the chemicals industry, is Stobaugh, 1968. For more general data, see Vernon, 1977, pp. 26-81.
5. On Japan, see Vogel, 1979, p. 72. On other countries, see Pinder, 1982a; a comparative analysis of national policies in different industries appears in various chapters in Pinder, 1982b.
6. U.S. government expenditures in 1983 were well above the national R&D expenditures from all sources in Japan, France, Germany, or the United Kingdom; see National Science Board, 1985, p. 187.
7. As of 1981 the proportion of the national market for manufactured imports affected by major nontariff barriers was higher in the United States than in six other industrial countries, including Japan. See Cline, 1984, p. 60.
8. A characteristic evaluation of this sort was that of Jean-Jacques Servan-Schreiber, 1968, p. 67.
9. The growth in the relative importance of merchandise imports in the United States is reflected in the increase in the ratio of nonagricultural nonpetroleum merchandise imports to value added by the U.S. manufacturing and mining sectors. That ratio rose from about 15 percent in the mid-1970s to about 23 percent in the early 1980s. From various issues of *Survey of Current Business* and *Economic Report of the President, 1985*, p. 244.
10. On steel, see Walters, 1983, pp. 484-486. On automobiles, see Cohen, 1983, p. 555.
11. Flows to the United States from abroad from royalties and fees, other private services, direct investments, and other private receipts rose from 1.3 percent of U.S. GNP in 1965 to 2.6 percent in 1984, reflecting a growth rate substantially higher than the

growth rate of total U.S. exports and of U.S. GNP. See *Survey of Current Business,* June 1985, pp. 40-41.

12. Stephen S. Cohen and John Zysman argue this position eloquently in the manuscript of their forthcoming book: *Manufacturing Matters: The Myth of the Post-Industrial Society,* New York: Basic Books.

13. The principal provision is Section 7 of the Clayton Act. See Sherman, 1978, pp. 38-41.

14. Between 1970 and 1984, the book value of all foreign direct investment in the United States rose from $13.3 billion to $159.6 billion. The manufacturing component in 1984 amounted to $50.7 billion. From various issues of the *Survey of Current Business.*

15. The joint-venture trend is described in Ohmae, 1985; and Mowery and Rosenberg, 1985.

16. The subject is explored in Crandall, 1981, especially pages 129-140.

17. Hard data on this subject are not easily available. One study, reflecting conditions in the latter 1970s, concludes that in the United States, business executives received compensation that was about 15 times that of an auto worker, whereas the comparable figure in Japan was 9 times, and in Sweden (after taxes) only a little over 2 times; the figures appear in an unpublished manuscript by Sidney Verba and Gary Orren, which in turn relies on various surveys by others to generate the comparison.

18. For an insightful summary of U.S. trade practices, see Grey, 1983.

19. For congressional reactions, see Ahearn and Reifman, 1984.

20. For illuminating discussions of the place of these codes, see Jackson, 1983; and Hufbauer, 1983.

21. The issue of nonsignatory rights has actually been in dispute in the GATT, with an ambiguous outcome; but the exclusionary intent in drafting the codes was reasonably clear. See Tarullo, 1984.

22. The significance of these departures from longstanding U.S. policy is emphasized in Samolis, 1984.

23. Although U.S. total manufacturing employment in 1983 was practically the same as in 1971, the number of union members in manufacturing declined by about 500,000, and the relative importance of the total union membership represented by manufacturing unions fell to 11.4 percent from 15.8 percent (U.S. Bureau of the Census, 1985, p. 423; *Economic Report of the President, 1985*, p. 275.)

REFERENCES

Abernathy, W., and R. H. Hayes. 1980. Managing our way to economic decline. Harvard Business Review (July-August):67-77.

Ahearn, R. J., and A. Reifman. 1984. Trade policymaking in the Congress. Pp. 36-66 in Recent Issues and Initiatives in U.S. Trade Policy, R. E. Baldwin, ed. Cambridge, Mass.: National Bureau of Economic Research.

Aho, M. C., and T. O. Bayard. 1980. American trade adjustment assistance after five years. The World Economy 3(November):359-376.

Bergsten, C. F., and W. R. Cline. 1985. The United States-Japan Economic Problem. Washington, D.C.: Institute for International Economics.

Bilzi, C. 1985. Recent United States trade arrangements: Implications for the most-favored nation principle and United States trade policy. Law and Policy in International Business 1:209-236.

Bozeman, B., and A. Link. 1985. Public support for private R&D: The case of the research tax credit. Journal of Policy Analysis and Management 4(3):376.

Brooks, H. 1985. Technology as a factor in U.S. competitiveness. Pp. 328-356 in U.S. Competitiveness in the World Economy, G. C. Lodge and B. R. Scott, eds. Boston: Harvard Business School Press.

Brooks, H. 1986. National science policy and technological innovation. Pp. 119-167 in The Positive Sum Strategy: Harnessing Technology for Economic Growth, R. Landau and N. Rosenberg, eds. Washington, D.C.: National Academy Press.

Casson, M., and G. Norman. 1983. Pricing and sourcing strategies in a multinational oligopoly. In The Growth of International Business, M. Casson, ed. London: George Allen and Unwin.

Cline, W. R. 1984. Exports of Manufactures from Developing Countries. Washington, D.C.: Brookings Institution.

Cohen, R. B. 1983. The prospects for trade and protectionism in the auto industry. In Trade Policy in the 1980s, W. R. Cline, ed. Cambridge, Mass.: MIT Press.

Crandall, R. W. 1981. The U.S. Steel Industry in Recurrent Crisis. Washington, D.C.: Brookings Institution.

Denison, E. F. 1979. Accounting for Slower Economic Growth. Washington, D.C.: Brookings Institution.

Diebold, J. 1983. The information technology industries: A case study of high technology trade. In Trade Policy in the 1980s, W. R. Cline, ed. Cambridge, Mass.: MIT Press.

Economic Report of the President. 1985. Washington, D.C.: U.S. Government Printing Office.

Freeman, C. 1982. The Economics of Industrial Innovation, 2nd ed. Cambridge, Mass.: MIT Press.

Gomez-Ibanez, J. A., R. A. Leone, and S. A. O'Connell. 1983. Restraining auto imports: Does anyone win? Journal of Policy Analysis and Management 2(2):196-219.

Grey, R. de C. 1983. A note on U.S. trade. Pp. 243-257 in Trade Policy in the 1980s, W. R. Cline, ed. Cambridge, Mass.: MIT Press.

Hexner, E. 1946. International Cartels. Chapel Hill: University of North Carolina Press.

Hufbauer, G. C. 1983. Subsidy issues after the Tokyo round. Pp. 327-361 in Trade Policy in the 1980s, W. R. Cline, ed. Cambridge, Mass.: MIT Press.

Jackson, J. H. 1983. GATT machinery and the Tokyo round agreements. Pp. 159-187 in Trade Policy in the 1980s, W. R. Cline, ed. Cambridge, Mass.: MIT Press.

Landes, D. 1969. The Unbound Prometheus. London: Cambridge University Press.

Lawrence, R. Z. 1984. Can America Compete? Washington, D.C.: Brookings Institution.

Lipsey, R. E. 1960. The theory of customs union. Economic Journal 70:496-513.

Lipsey, R. E. 1984. Recent Trends in U.S. Trade and Investment. Report No. 565. Cambridge, Mass.: National Bureau of Economic Research.

Lipsey, R. E., and I. B. Kravis. 1985. The competitive position of U.S. manufacturing firms. Banco Nazionale di Lavoro Quarterly Review (June):127-154.

Lodge, G. 1984. The American Disease. New York: Alfred A. Knopf.

Magaziner, I. C., and R. B. Reich. 1982. Minding America's Business. New York: Harcourt Brace Jovanovich.

Mansfield, E. 1968. The Economics of Technological Change. New York: W. W. Norton.

Mowery, D., and N. Rosenberg. 1985. Commercial aircraft: Cooperation and competition between the U.S. and Japan. California Management Review (Summer):70-92.

National Science Board. 1985. Science Indicators 1985. Washington, D.C.: National Science Board.

Nelson, R., ed. 1982. Government and Technical Progress. New York: Pergamon Press.

Norton, R. D. 1986. Industrial policy and American renewal. Journal of Economic Literature 24(1):12-17.

Ohmae, K. 1985. Triad Power: The Coming Shape of Global Competition. New York: Free Press.

Pastor, R. A. 1980. Congress and the Politics of U.S. Foreign Economic Policy 1929-1976. Berkeley: University of California Press.

Patterson, G. 1983. The European Community as a threat to the system. Pp. 223-241 in Trade Policy in the 1980s, W. R. Cline, ed. Cambridge, Mass.: MIT Press.

Pavitt, K. 1971. The Conditions for Success in Technological Innovation. Paris: Organization for Economic Cooperation and Development.

Pinder, J. 1982a. Causes and kinds of industrial policy. Pp. 44-51 in National Industrial Strategies and the World Economy, J. Pinder, ed. Totowa, N.J.: Allanheld, Osmun and Company.

Pinder, J., ed. 1982b. National Industrial Strategies and the World Economy. Totowa, N.J.: Allanheld, Osmun and Company.

Porter, M. 1985. Competitive America. New York: Free Press.

President's Commission on Industrial Competitiveness. 1985. Global Competition, The New Reality: The Report of the President's Commission on Industrial Competitiveness, Vols. 1 and 2. Washington, D.C.: U.S. Government Printing Office.

Reich, R. 1983. The Next American Frontier. New York: Penguin Books.

Richardson, J. D. 1983. Worker adjustment to U.S. international trade: Programs and prospects. Pp. 393-417 in Trade Policy in the 1980s, W. R. Cline, ed. Cambridge, Mass.: MIT Press.

Roy, A. D. 1982. Labor productivity in 1980: An international comparison. National Institute Economic Review 101(August):26-37.

Samolis, F. B. 1984. SOS for the CBI: Lessons from the Caribbean basin initiative. Pp. 137-142 in Managing Trade Relations in the 1980s. Totowa, N.J.: Rowman and Allanheld.

Saxonhouse, G. 1983. The micro- and macroeconomics of foreign sales to Japan. Pp. 259-263 in Trade Policy in the 1980s, W. R. Cline, ed. Cambridge, Mass.: MIT Press.

Schultze, C. L. 1973. The economic conquest of national security policy. Foreign Affairs (April):526.

Servan-Schreiber, J-J. 1968. The American Challenge. New York: Atheneum.

Sherman, R. 1978. Antitrust Policies and Issues. Reading, Mass.: Addison-Wesley.

Stobaugh, R. B. 1968. The product-life cycle, U.S. exports and international investment. D.B.A. thesis. Harvard Business School.

Tarullo, D. K. 1984. The MTN subsidies code: Agreement without consensus. Pp. 63-99 in Emerging Standards of International Trade and Investment, S. J. Rubin and G. C. Hufbauer, eds. Totowa, N.J.: Rowman and Allanheld.

Teece, D. J. 1983. Technological and organization factors in the theory of the multinational enterprise. Pp. 54-62 in The Growth of International Business, M. Casson, ed. London: George Allen and Unwin.

Thurow, L. 1980. The Zero-Sum Society. New York: Basic Books.

U.S. Bureau of the Census. 1985. Statistical Abstract of the United States: 1985. 105th edition. Washington, D.C.: U.S. Government Printing Office.

United Nations. 1983. Yearbook of International Trade Statistics. New York: United Nations.

Verba, S., and G. Orren. 1985. Equality in America. Cambridge, Mass.: Harvard University Press.

Vernon, R. 1955. Foreign trade and national defense. Foreign Affairs (October):77-88.

Vernon, R. 1977. Storm Over the Multinationals. Cambridge, Mass.: Harvard University Press.

Vernon, R., and W. H. Davidson. 1979. Foreign Production of Technology-Intensive Products by U.S.-Based Multinational Enterprises. Report to the National Science Foundation, No. PB80 148638, January. Washington, D.C.: National Science Foundation.

Vogel, E. 1979. Japan as No. 1: Lessons for America. Cambridge, Mass.: Harvard University Press.

Walters, I. 1983. Structural adjustment and trade policy in the international steel industry. In Trade Policy in the 1980s, W. R. Cline, ed. Cambridge, Mass.: MIT Press.

Does Technology Policy Matter?

HENRY ERGAS

How do technology policies differ among nations? What impact do these differences have on innovation and, more generally, on industrial strutures? These questions are the central concerns of this chapter.

Innovation is the use of human, technical, and financial *resources* to find a way of doing things. As an inherently uncertain process, it requires *experimentation* with alternative approaches, many of which may prove unsuccessful. Even fewer will survive the test of *diffusion*, where ultimate economic returns are determined. The historical success of the capitalist system as an engine of growth arises from its superiority at each of these levels: generating the resources required for innovation, allowing the freedom to experiment with alternative approaches, and providing the incentives to do so.[1]

Though relying primarily on market forces, the system has interacted with government at two essential levels. The first relates to the harnessing of technological power for public purposes. Nation-states have long been major consumers of new products, particularly for military uses, and the need to compete against other nation-states provided an important early rationale for strengthening national technological capabilities. Whether this rationale persists as the primary motive for government action is a major factor shaping each country's technological policies (Earle, 1986).

The second arises from the system's dependence on its social context. The development and diffusion of advanced technologies requires a system of education and training as a basis for supplying technology and skills, a legal framework for defining and enforcing property rights, and processes such as standardization to reduce transactions costs and increase the transparency and efficiency of markets. These are, at least in part, public goods. The benefits of investment in education are appropriated by a multitude of economic actors, and those of the system of property rights are even more widely spread.[2] The way these public goods are provided, and the role industry plays in this respect, differs greatly from country to country.

This chapter examines the interactions between the technological system and government policy in seven industrialized countries: the United States, the United Kingdom, France, Germany, Switzerland, Sweden, and Japan. It pays particular attention to the relation between innovation policy and industrial structures. The countries examined are placed in three groups.

Technology policy in the United States, the United Kingdom, and France remains intimately linked to objectives of national sovereignty. Best described as "mission-oriented," the technology policies of these nations focus on radical innovations needed to achieve clearly set out goals of national importance. In these countries, the provision of innovation-related public goods is only a secondary concern of technology policy.

In contrast, technology policy in Germany, Switzerland, and Sweden is primarily "diffusion-oriented." Closely bound up with the provision of public goods, the principal purpose of these policies is to diffuse technological capabilities throughout the industrial structure, thus facilitating the ongoing and mainly incremental adaptation to change. Finally, Japan is in a group of its own. Its technology policy is both mission-oriented *and* diffusion-oriented, and the form the policy takes differs in important respects from that in the other countries.

Every taxonomy involves a loss of information, and the one proposed here does not escape this general rule. Thus, the United States has important policies—for example, in agriculture and in medical research—that are diffusion-oriented; equally, Germany and Sweden have major mission-oriented programs. But the focus of policy does differ in the three groups, and this allows a clearer examination of the relation between technology policy and innovation performance.

These differences in policy stance—though not as sharp as they may at first appear to be—are important in shaping patterns of technological evolution, but the central hypothesis of this chapter is that technology policies are a *facilitating* rather than *explanatory* factor. The critical variables lie in how industry responds to the results and signals of efforts to upgrade national technological capabilities. In turn, this depends to a substantial extent on the environment in which industry operates. Technology policies cannot, in other words, be assessed independently from their broader economic and institutional context.

A central feature of this context is a country's *technological infrastruc-*

ture—its system of education and training, its public and private research laboratories, its network of scientific and technological associations. The effectiveness of this infrastructure depends not only on its internal functioning but also on the way a country's factor and product markets respond to innovation opportunities.

Overall, this suggests that, even within the framework of a market economy, the process by which innovations are generated, selected, and imitated will differ according to the features of each country's institutional and economic structure. In exploring these features and their relation to countries' technology policies, this chapter follows the broad grouping set out above: The next three sections examine, respectively, the technology policies of the mission-oriented countries, namely, the United States, the United Kingdom, and France; the diffusion-oriented countries, namely, Germany, Switzerland, and Sweden; and Japan. The last two sections present, respectively, a synthesis of similarities and differences, with analyses of their broader implications for economic performance, and conclusions for policy formulation.

THE MISSION-ORIENTED COUNTRIES

Mission-oriented research can be described as big science deployed to meet big problems (Weinberg, 1967). It is of primary relevance to countries engaged in the search for international strategic leadership, and the countries in which it dominates are those where defense accounts for a high share of government expenditure on R&D (Table 1). Though it has also been used in these countries to meet perceived technological needs in civilian markets (for example, in nuclear energy or telecommunications), the link to national sovereignty provides its major rationale.

The dominant feature of mission-oriented R&D is *concentration*. First and most visibly, this refers to the centralization of decision making. As

TABLE 1 Share of Defense-Related R&D in Total Government Expenditure on R&D, 1981

Country	Percent Defense-Related
United States	54
United Kingdom	49
France	39
Sweden	15
Switzerland	12
Germany	9
Japan	2

SOURCE: Organization for Economic Cooperation and Development.

its name implies, the goals of mission-oriented R&D are centrally decided and clearly set out, generally in terms of complex systems meeting the needs of a particular government agency. Specifying these needs and supervising project implementation concentrates a considerable amount of discretionary power in the hands of the major funding agencies.

Concentration also extends to the range of technologies covered. Virtually by its nature, mission-oriented research focuses on a small number of technologies of particular strategic importance—primarily in aerospace, electronics, and nuclear energy. As a result, government R&D funding in these countries is heavily biased toward a few industries that are generally considered to be in the early stages of the technology life cycle (Table 2).

The scale of mission-oriented efforts also limits the number of projects and restricts the number of participants. At any particular time, only a small share of each country's firms, likely among the larger ones, will have the technical and managerial resources required to participate in these programs. The concentration of government R&D subsidies on a small number of large firms is therefore also a feature of the countries in this group.

Overall, mission-oriented programs concentrate decision making, implementation, and evaluation. A few bets are placed on a small number of races; but together, these bets are large enough to account for a high share of each country's total technology development program. This concentration raises two obvious questions: First, how successful are the bets in relation to their own objectives? And second, do they have any effect on the efficiency with which the many other races are run—that is, are

TABLE 2 Proportion of Total National Public R&D Funding by Type of Industry, 1980 Estimates

	Percentage of Total Public R&D Funding		
Country	High-Intensity Industry	Medium-Intensity Industry	Low-Intensity Industry
United States	88	8	4
France	91	7	2
United Kingdom	95	3	2
Germany	67	23	10
Sweden	71	20	9
Japan	21	12	67

NOTE: High-, medium-, and low-intensity R&D industries are defined as firms whose ratios of R&D expenditures to sales are, respectively, more than twice, between twice and half, and less than half the manufacturing average.
SOURCE: Organization for Economic Cooperation and Development.

technological capabilities more broadly diffused through the industrial structure? These questions will be considered in turn.

Direct Effectiveness

Attempting cost-benefit analyses of major mission-oriented programs involves enormous difficulties (Hitch and McKean, 1960). Three criteria for evaluating success can nevertheless be established: First, are stated product development goals being met? Second, is this being done within the original limits of time and cost? And third, are objectives for commercial markets being achieved?

No country's programs perform extremely well when measured against these criteria. On balance, the effort in the United Kingdom has probably been the least successful, whereas that in France and the United States has generated a mixed record. Three factors seem to be critical in differentiating success from failure. First, do the agencies involved have the technical expertise, financial resources, and operating autonomy required to design and implement the program—and the incentives to ensure that it succeeds? Second, are relations with outside suppliers such as to provide appropriate incentives and penalties and do they allow for experimentation with alternative design approaches? Third, can agencies be prevented from expanding their "missions" indefinitely and, in particular, from moving into areas for which their capabilities and structures are inappropriate?

The answers to these questions have differed in each of the three mission-oriented countries considered in this section.

United Kingdom

The United Kingdom's major difficulties arise from the pervasive lack of incentives in its system of mission-oriented R&D.[3] The British system of public administration—with its emphasis on anonymity, committee decision making, and administrative secrecy—ensures that individual public servants have little interest in "rocking the boat." The emphasis on internal and procedural accountability also makes government reluctant to devolve major projects to reasonably autonomous entities, so that responsibilities are tangled, decision making is cumbersome, and the organizational and cultural context is inappropriate for developing new technologies. At the same time, the propensity of British agencies to form "clubs" with their suppliers—within which each supplier is treated on the basis of administrative equity rather than commercial efficiency—weakens whatever incentives suppliers may have to seek an early lead, while also ensuring that the resources available are so thinly spread as to be ineffectual. Finally, the reluctance to build penalty clauses into development contracts, and to terminate unsuccessful projects (particularly when this would jeopardize the viability of an indigenous supplier), aggravates an inherent tendency to cost overruns.

France

France's relative success arises in considerable part from the great political legitimacy, operating autonomy, and technical expertise of its user agencies, combined with the strong incentives for success built into the highly personalized nature of power and careers in the French public administration.[4] Particularly over the last decade, there has also been an effort to increase the competitive pressures bearing on suppliers, notably through tighter controls on costs, recourse to penalty clauses, and easing previous market-sharing arrangements. The effects of these moves have been heightened by improved financial and operating control within the agencies themselves.

However, the French system has two major weaknesses. First, resource constraints have usually prevented experimentation with alternative design approaches, and the number of suppliers involved in each major project has typically been small.[5] Second, though the French system has been compared favorably to that of the United Kingdom because there has been a reasonable willingness to run down (if not terminate) failures, the system has been highly vulnerable to goal displacement as a sequel to success. Agencies that have successfully accomplished a mission perpetuate themselves by designing new missions, frequently in areas unrelated to their original function. This "Frankenstein" effect is particularly noticeable in the energy and communications fields, where agencies have sought to expand their power base by diversifying their operations, generally into markets for which their technological capabilities and organizational structures are inappropriate. As a result, success in one period has in several cases been followed by failure in the next; and the system has had few mechanisms for reallocating resources smoothly.[6]

United States

Considering only the efficiency with which projects are designed and implemented, the United States is intermediate between the United Kingdom and France; but it has over them the great advantage of scale.[7] This advantage has three important dimensions. First, U.S. agencies draw on a much larger pool of external technological expertise both in selecting and implementing projects—and have much better mechanisms for doing so, notably in university research. Second, funding for mission-oriented programs in the United States, particularly in defense, rarely falls short of the critical mass required to complete the development stage and usually has a higher continuity than program funding elsewhere. Third, the scale of funding is large, and the range of qualified suppliers is wide. Even the relatively small sums spent by the U.S. Department of Defense on programs of the Defense Advanced Research Projects Agency are large in relation to total defense R&D in the United Kingdom and France. The result is that experimentation almost invariably occurs with alternative design approaches and philosophies, even if only in the early steps of program conception.

The United States may also benefit from the high degree of accountability inherent in its system of congressional scrutiny. This system has generated strong pressures for terminating unsuccessful projects, notably in the civilian sector (the supersonic transport plane and synfuels being prime examples), but seems to exercise much less control on the defense sector. Thus, an incidental effect of the system is that military programs may be allowed to continue too long, and some largely civilian programs are shut down too early. It has been argued that this places an excessive burden of financing projects with a high "public goods" content on the private sector. The safety and decommissioning of nuclear power plants may be cases in point (Brooks, 1983).

Any overall assessment of the direct effectiveness of mission-oriented research must therefore be mixed; but the immediate returns on the research do appear to be higher in the United States and France than in the United Kingdom. However, even in the United States the products conceived directly by mission-oriented programs account for only a small share of the economy (Riche et al., 1983); the extent to which technology generated in these programs spreads to other areas of activity is therefore a major component of its overall impact.

Secondary Effectiveness

There are relatively few studies of the extent of secondary effects of mission-oriented technology policies or of the pace at which such effects occur. The few studies that do exist come to widely differing conclusions, frequently reflecting individual authors' views of the desirability of defense spending. None of the studies draws international comparisons. Two broad statements can nonetheless be advanced on the basis of the existing material: first, in every country, the direct spin-offs—in the sense of immediate commerical use of the results of mission-oriented research—are limited;[8] second, the indirect spin-offs—arising mainly from improvements in skills and in technical knowledge transferable from the mission-oriented environment to that of commercial competition—appear to occur both in greater number and more rapidly in the United States than in the United Kingdom or France.[9]

It can be argued that the greater number and frequency of indirect spin-offs in the United States are partly due to differences in the way programs are designed and implemented. But the impact of these differences is compounded by differences in the countries' economic structures and scientific and technological environments.

The Role of Program Design

Four factors distinguish the design and implementation of mission-oriented programs in the United States from that of their counterparts in the United Kingdom and France. The first is the more limited direct role of the public sector in mission-oriented R&D in the United States. In general, the U.S. government performs a small share of its research in-house; the bulk of it is contracted to outside sources (Table 3). Even the management of national laboratories has been separated to a considerable extent from the public sector and devolved to universities or to private companies. Problems of technology transfer from the public to the private sector therefore concern a smaller share of government-funded R&D than is the case in France or the United Kingdom.

Second, mission-oriented research in the United States involves a greater number and diversity of agents. It is true that within the private sector, most government research and procurement contracts go to a small number of suppliers. But the sums flowing to university research and to small and medium-size businesses are large in absolute terms.[10] Thus, the number of small firms receiving 20 percent or more of their total R&D finance from government sources is nearly 10 times larger in the United States than in the United Kingdom or France. Moreover, insistence in defense

TABLE 3   Share of Government-Financed R&D Performed in the Government Sector

Country (Year)	Share Performed by Government
France (1983)	46.8
United Kingdom (1981)	38.9
Federal Republic of Germany (1981)	31.6
United States (1983)	25.7
Switzerland (1981)	24.7

SOURCE: Organization for Economic Cooperation and Development.

procurement on "second-sourcing" of key components ensures a fairly broad diffusion of technological capabilities.

The effects of this dispersion are compounded by a third factor, namely, greater U.S. willingness to disseminate the results of mission-oriented programs.[11] Despite obvious security concerns, U.S. defense R&D programs have generally either made their results public or at least made them known to a wider circle than that immediately involved in the program. The information inherent in these results—such as measurement standards, properties of materials, or even identification of unsuccessful approaches to technical problems—is an important "public good."

A greater U.S. willingness to disseminate results probably contains an element of bowing to the inevitable: Given the number and range of participants, results will be known sooner or later. However, other factors have also been at work. The widespread dissemination of results has been important in securing ongoing political approval for the programs. It has also been a way of preventing contractors from consolidating a "first-mover advantage" over competitors. At universities especially, dissemination has been facilitated by a research community that generally has not questioned the legitimacy of the program so long as their results could be fed into the system of "public or perish!"

The dissemination of the results of mission-oriented programs in the United Kingdom and France differs from that in the United States in three respects. First, after programs are set up and running, there is little external political pressure to disseminate results. Second, the members of the program "club" themselves have little interest in seeing results publicized and tend to count more heavily in decisions about dissemination. Third, the external environment—notably that in the universities—has been perceived as probably hostile and possibly untrustworthy. As a result, the information generated by mission-oriented programs has tended to remain confined to a small circle of participants.

Finally, the U.S. government moved somewhat earlier than its counterparts in France and the United Kingdom to encourage commercialization of the results of government-financed R&D. The National Aeronautics and Space Administration and a few other federal agencies have long had specified units concerned with technology transfer. Regarding government-financed R&D in the private sector, the 1980 Patent Law Amendments Act established a uniform policy allowing contractors—notably, small businesses, universities, and nonprofit laboratories—to own inventions resulting from federal R&D funding. The assurance this act provides of clear title to government-funded inventions has greatly facilitated patent licensing by universities and other federal contractors to industry and has encouraged industrial participation in federally supported university research.

TABLE 4   Research Scientists and Engineers in the
Labor Force, 1981

Country	Number per 1,000 of Labor Force
United States	6.2
Japan	5.4
Federal Republic of Germany	4.7
United Kingdom	3.9
Norway	3.8
France	3.6

SOURCE: Organization for Economic Cooperation and Development.

Differences in the Environment

Economic interests in the United States therefore have greater direct or indirect access to whatever may be transferable in the outcomes of mission-oriented programs. At the same time they are well placed to exploit these results for commercial purposes and have substantial incentives to do so.

Lower Degree of Crowding Out The sheer size of the U.S. scientific and technological system means that mission-oriented programs probably "crowd out" other research efforts only to a limited extent. The size differential is particularly marked in terms of the stock and flow of research manpower. The share of R&D scientists and engineers in the U.S. labor force is one-third greater than that in the United Kingdom and France (Table 4). The share of secondary students going on to university training in the United States is about double that in France or the United Kingdom (Table 5), and the proportion of those students choosing scientific or engineering training is reasonably responsive to market circumstances.[12] To this difference in endowment must be added the effect of inflows of scientists and engineers from overseas. In 1982, foreign-born scientists and engineers accounted for fully 17 percent of all scientists and engineers employed in the United States.

Accessibility and Mobility of Scientific Know-how The U.S. stock of human and technological capital, in addition to being relatively abundant is also more easily accessible. It is, in the first place, accessible through contract research, both with private research firms and with universities. Though the share of university research financed by industry in the United States is not high, the links between universities and industry have traditionally been strong (Noble, 1977; Ben-David, 1968)—far stronger, at least, than in France or the United Kingdom, both these countries lagging even by

European standards in this respect (Ahlström, 1982; Organization for Economic Cooperation and Development, 1984a; Ben-David, 1968). These links take several forms: active efforts by U.S. universities to commercialize their technological skills, widespread consulting for industry by university scientists and engineers, frequent coauthorship of journal articles by researchers in industry and academia, and sizeable gifts of equipment by industry to university research facilities.

The operation of the U.S. labor market also promotes the accessibility of its stock of human and technological capital. In general, the U.S. labor force is more mobile between employers and regions than the labor force in Europe: Average job tenure is about 20 percent lower in the United States than in France or the United Kingdom; the share of the labor force crossing regional boundaries each year is—at about 3 percent—double that in Europe. Moreover, U.S. scientists and engineers are almost as mobile as other segments of the labor force: Their average job tenure is only about 15 percent higher than the average. In contrast, mean tenure in France with a given employer is nearly 40 percent higher for highly qualified staff than for the labor force as a whole (Pham-Khac and Pigelet, 1979; Stevens, 1986).

Differences in labor mobility are even greater regarding movement from university to industry. Some 2 to 3 percent of all U.S. scientists and engineers move from academia to industry or vice versa every year; the figure for France can be estimated at well below 0.5 percent.[13] The civil service status of public sector researchers in France makes movement difficult and eliminates incentives to move.

TABLE 5   Diplomas Giving Access to Higher Education as Proportion of Age Group

Country (Year)	Percent
Japan (1981)	87
Sweden (1982)	82
United States (1980)	72
Federal Republic of Germany (1982)	26
Denmark (1980)	25
France (1983)	28
Italy (1981)	39
United Kingdom (1981)	26
Finland (1980)	38
Austria (1978)	13
Netherlands (1981)	44

SOURCE: Organization for Economic Cooperation and Development.

Competition in Factor and Product Markets High levels of mobility of scientists and engineers in the United States ensure that technological capabilities generated by mission-oriented research are rapidly diffused among firms but do not ensure that such capabilities will rapidly be exploited. This in turn hinges on the intensity of competition in product markets, which encourages firms to innovate. Three factors distinguish the United States in this respect: the receptiveness of capital markets to innovation efforts, the extent of the threat of new firm entry, and the incentives to innovation arising from a large and unified market.

Capital markets in the United States are distinguished from those elsewhere largely by two features: the depth and breadth of equity markets and the availability of venture capital finance for start-up companies (Gönenç, 1986). It can be argued whether these institutions have proved appropriate for financing long-term market share strategies; but—perhaps because they provide a low-cost means for realizing capital gains—they appear to do reasonably well at providing concurrent finance for a broad range of innovation efforts. Certainly the balance of evidence indicates that they are effective mechanisms for the monitoring and diversification of innovation-related risks and opportunities.

The functioning of capital markets reinforces the degree of competition in U.S. product markets in two important respects. First, the widespread availability of venture capital—together with a range of other environmental factors that reduce the costs of setting up and dissolving businesses—increases the threat of entry by new companies. This is reflected in rates of creation and disappearance of new manufacturing firms, which are nearly twice those in France (Arocena, 1983; Ergas, 1984b). Ideas not exploited by large companies are likely to be tried out quickly by an entrepreneur. This is of particular importance in the early stages of a new technology, when a large number of alternative design approaches are being explored (Clark, 1985; Freeman, 1974; Nelson and Winter, 1982).

Second, an active market for corporate control provides an effective means of liquidating new firms that do poorly and incorporating into larger concerns the activities of those that do well. At the same time, the takeover market reduces the risks associated with entry by diversification. Large U.S. firms have tended to enter new markets by buying smaller firms already operating in those markets, knowing that if the venture failed, it could be disposed of (Scherer and Ravenscraft, 1984).

The effects of potential competition are compounded by the far greater supply in the United States of potential entrants into advanced technology markets. More than 15,000 firms in the United States have R&D laboratories; this compares with about 1,500 in France and 800 in the United Kingdom. The number of firms with some technological capability in any given area is likely to reflect this differential. This provides the United

States with a large seedbed capable of responding quickly to the "focusing" effects of innovations and acting as an incubator for potential entrepreneurs. It also provides a large number of firms capable of acting as a "fast second"—moving into a new market as its attractiveness is established and as the appropriate technological approach becomes clear.

Size of the U.S. Market The nature of competition in the U.S. market also intensifies firms' interests in new product areas, notably as a technology approaches the stage where mass marketing becomes essential. Three factors are of particular relevance. First, because of the importance of economies of scale in a relatively homogeneous market, firms vie for leadership in the transition to mass production and marketing.[14] Second, reliance on de facto or proprietary standards provides the firm whose product emerges as a dominant design with a considerable advantage. Third, the U.S. market appears to be highly sensitive to "perceptual" product differentiation, which tends to favor early entrants to the mass marketing and production stage.[15]

Each of these factors can create first-mover advantages, compounding the benefits the United States derives from having a greater number of potential first-movers. As a result, the two basic components of the "swarming" process—by which firms flock to an emerging market—tend to operate particularly rapidly in the United States: the *experimentation* stage, in which a range of alternative design approaches is explored, frequently by smaller firms; and the transition to *mass commercialization*, as the technology matures to the point of market acceptability. Preeminence in both of these stages increases the likelihood that U.S. firms will be well placed to spot an emerging dominant design.

The Link to Performance

The preceding discussion of mission-oriented countries can be summarized as follows. In the United Kingdom, mission-oriented research has tended to yield few direct benefits while possibly crowding out a substantial share of commercial R&D. The indirect spin-offs have been low, creating a "sheltered workshop" type of economy: a small number of more or less directly subsidized high-technology firms, heavily dependent on and oriented to public procurement, and a traditional sector that draws little benefit from the high overall level of expenditure on R&D.[16]

In France, mission-oriented research efforts have themselves been reasonably successful. This has created export markets for France, notably in the largest weapons-importing countries of the Third World and in other countries where state-to-state trading is important. However, the spin-offs from these efforts have been relatively limited, so that French industry has become increasingly dualistic in its access to, and reliance on, advanced technology. This has been most visible in France's shifting pattern of international trade. Exports of products requiring a high intensity of skills, though rising, have concentrated to a growing extent on Third World markets, reflecting the predominance of state-to-state trading, whereas in trade with the advanced countries, the relative skill intensity of French exports has tended to diminish. The centralized and concentrated nature of mission-oriented research has therefore led to an increasingly polarized pattern of specialization.[17]

The situation of the United States is more complex. Although the direct effectiveness of mission-oriented programs is no higher than in France, the results of these programs tend to diffuse particularly rapidly through the U.S. economy. This rapid diffusion is a result of three features: the wide range of economic interests capable of exploiting these results for commercial purposes, the low level of the obstacles they encounter in seeking to do so, and the strength of the incentives for rapid exploitation. The mission-oriented stage of research in the United States remains highly centralized, but its results are more rapidly carried over into the decentralized experimentation of the commercial market.

Particularly in recent years, the rapid carryover of the results of the U.S. programs has generated advantages that may be cumulative at the level of the firm but are not cumulative at the level of the product. More specifically, although U.S. firms appear to retain many of their established strengths, U.S. production sites have proved considerably better at the experimentation stage than the follow-on to mass production (Lipsey and Kravis, 1985).[18] This partly reflects the macroeconomic circumstances associated with the overvaluation of the dollar, but more fundamental factors may also be at work.

Historically, the United States has lacked a system for training craftsmen, while possessing an abundance of higher-skilled (white collar) and lower-skilled or unskilled workers (Floud, 1984; National Manpower Council, 1954; Floud, 1984). At the same time, the structure of blue-collar earnings in the unionized parts of U.S. industry (with low differentials between trainee wages and those of craftsmen) and high labor mobility have discouraged employer investment in transferable skills (Glover, 1974; Mitchell, 1977; Ryan, 1984). Combined with a large and unified national market, this pushed U.S. manufacturing firms in two directions: pioneering mass production techniques that made little use of craft labor and developing organizational innovations intensive in their use of managerial or supervisory staff—such as multiplant production, multidivisional management, and the multinational firm.

The advantage that superior mass production techniques gave U.S.

production sites has tended to erode over time, for at least three reasons. First, in an increasingly integrated world economy, being located in the world's largest single market is of diminishing importance as a determinant of competitiveness. Second, the quality of the U.S. labor force—and particularly that part with only a high school degree or less—has probably declined relative to that overseas, and notably relative to that in Japan (Murray, 1984, pp. 96-112). Third, classical mass production techniques along "Taylorist" lines may be of diminishing effectiveness as the variability and differentiation of products increases, as product workmanship becomes a more important factor in consumer choice, and as new technologies for "mid-scale" production become available (Ergas, 1984a).

These factors place the U.S. manufacturing industry at a clear disadvantage, but they have less impact, if any, on the service sector. As a result, U.S. firms tend to reap the advantages of innovative capabilities in manufacturing mainly at the early stages of the product life cycle (or, if the dollar is low enough, in products that are mature). In services, the gains from innovation have been consolidated further downstream as markets grow. Given a reasonably flexible and open economy, this pattern is reflected in the structure of trade. Thus, resources have tended to cluster around emerging or science-based industries.

In this sense, the United States comes closest to the classical product cycle model, abandoning mature industries in favor of activities with better growth prospects.[19] A system of mission-oriented research, which helps ensure that the frontiers of these activities are constantly being explored, may provide a useful source of ongoing stimulus to this process. It therefore has a certain degree of coherence relative to the U.S. economy. Whether this process would not occur of its own volition—that is, even in the absence of mission-oriented research—remains an open question.

THE DIFFUSION-ORIENTED COUNTRIES

Diffusion-oriented policies seek to provide a broadly based capacity for adjusting to technological change throughout the industrial structure. They are characteristic of open economies where small and medium-size manufacturing enterprises remain an important economic and political force and where the state, bearing the interests of these firms in mind, aims at facilitating change rather than directing it.[20]

The primary feature of these policies is *decentralization*. Specific technological objectives are rarely set at a central level. Central government agencies play a limited role in implementation, preferring to delegate this stage either to industry associations or to cooperative research organizations dominated by industry. Whatever funds are disbursed tend to be fairly widely spread across firms and industries, with the high-technology industry obtaining a far lower share than in the mission-oriented countries.

Given this degree of decentralization, the precise boundaries of technology policy are often difficult to identify. Switzerland, for example, would certainly deny having a "technology policy" in the sense in which France has one. A more fruitful approach is to view technology policy in these countries as an intrinsic part of the provision of innovation-related public goods: notably education, product standardization, and cooperative research. These countries' distinguishing feature is the importance they attach to the organization and high quality of the provision of these goods and the decentralized mechanisms they have developed for supplying them.

The Economic and Institutional Framework

The priority accorded to the provision of public goods has its origin in process of industrialization in these countries. Two interrelated features distinguished this process: an emphasis on "education push," notably through innovations in higher education and in the training of engineers (Ahlström, 1982), and an early specialization in the chemical and electrical industries on the one hand and in mechanical engineering on the other.[21] This early pattern of specialization fed back into the demand for innovation-related public goods.

The chemicals and electrical industries were distinguished from the start by the closeness of their links to the science base (Beer, 1959; Freeman, 1974; Liebenau, 1985; Rosenberg, 1976; Rosenberg and Birdzell, 1986). They needed a high-quality university system, capable of training scientists for industry, of monitoring scientific developments worldwide, of providing external support to the emerging industrial research laboratories. Achieving this system in turn depended on developing an increasingly efficient and effective school system, which could prepare and select candidates for higher education. The Lutheran tradition of universal literacy and broadly based instruction provided an ideal basis for this evolution (Sandberg, 1979).

The chemicals and electrical industries therefore acted as a politically powerful and well-organized lobby for education and for academic research. Being highly concentrated and largely cartellized, they were fully capable of mobilizing in their collective interest (Forman, 1974; Schröder-Gudehus, 1972). But the needs of the mechanical engineering industries were different. First, whereas chemicals and electricals were science-based, mechanical engineering relied on learning-by-doing and on the tacit, unformalized know-how of skilled craftsmen. Second, whereas chemicals and electricals tended to be concentrated, mechanical engineering was not, mainly because a high level of decentralization was more efficient in monitoring the type of team production required to maintain the quality of workmanship.

For decentralization to persist, the engineering industry had to resolve three major problems. First, it had to be able to draw on an external pool of skilled labor, since no single small or medium-sized firm could efficiently rely on its internal labor market alone. Second, it had to reduce the transaction costs involved in the decentralized production of components that are close complements from an economic viewpoint—e.g., nuts and bolts. Third, it had to find ways of keeping firms up to date with technological developments, ensuring that the fruits of technical advance accumulated and were appropriated at the level of the industry as a whole, rather than primarily or solely at the level of the firm.

Mechanical engineering was therefore an active lobby for three policies: comprehensive vocational education, product standardization, and cooperative research. It sought these policies mainly through provision by industry associations rather than by government; and, particularly in Germany and Switzerland, this coincided with a governmental practice of according quasi-public status and functions to private bodies, originally to regulate markets (Berger, 1981; Katzenstein, 1985a).

As it has evolved in these countries, the overall system of public policy affecting technological capabilities has therefore had three key features.

Vocational Education

The most significant feature is probably the depth and breadth of investment in human capital, centering on the dual system of education. This involves comprehensive secondary education based on streaming into a high-quality university system that is paralleled by an extensive system of vocational education.[22]

A distinguishing characteristic of the educational component of the system in diffusion-oriented countries is high retention rates. More than 85 percent of 17-year-olds are in the education and training system in these countries; this compares with around 60 percent in the United Kingdom and 70 percent in France. The system is characterized further by a relatively high level of per capita expenditure on education at all levels. Over the last decade, the elasticity of total public educational expenditure with respect to gross domestic product (GDP) has been around 5 times higher in Switzerland than in the United States, starting from a base where Swiss expenditure per pupil was already a higher share of per capita GDP. Finally, the system is notable for its far-reaching certification. Only some 10-15 percent of the age cohort leave school with no certificate or qualification whatsoever, compared with 20 percent in the United States and as much as 40 percent in France and the United Kingdom.

Particularly in the German-speaking countries, the skill certification of large parts of the youth cohort occurs through the system of apprenticeship-based vocational education. More than 50 percent of 17-year-olds in Germany and Switzerland are enrolled in apprenticeships, compared with about 10 percent in France and the United Kingdom. These high rates of participation are encouraged both by a substantial differential between trainee wages and those of skilled craftsmen (Jones and Holenstein, 1983) and by a well-organized and extensive system for training apprentices. Thus, apprenticeships are highly structured programs of several years' duration. They include a combination of enterprise training and college education and culminate in standardized formal examinations. Moreover, completion of apprenticeships is only one stage in skill training: The classification of examination-certified vocational skills forms a continuum from the craftsman to the most highly trained engineer, and movement along this continuum is a relatively standard feature of working life.[23]

There is a high level of industry involvement throughout this system. In the general education sector, the main links are between industries and universities (these will be discussed below). But the core of industry involvement is in vocational education. The apprenticeship system is jointly financed and controlled by employers (acting mainly through industry associations) and local education authorities, with trade unions also providing an important input. Industry associations play a major role in defining and revising curricula and in monitoring the system's effectiveness. Combined with the emphasis on formal, written examinations, this ensures that the skills acquired are highly transferable between employers and can be adapted to improvements in the industry's technology base.

Overall, this structure of investment in human capital yields two outcomes: a university system capable of keeping up with the frontiers of science, though not necessarily pioneering their exploration, and a very high level of intermediate skills in the working population.[24] The fact that these skills are certified through a standardized system of examinations erodes the advantages that internal labor markets would otherwise have had in information about individual workers' skills, and hence tends to favor smaller firms. In turn, the ongoing nature of certification encourages relatively high levels of mobility for skilled craftsmen with work experience, providing a further channel for the interfirm diffusion of technology (Glover and Lawrence, 1976; Maurice et al., 1982; Office Fédéral de l'Industrie, des Arts et Métiers et du Travail, 1980).

Industrial Standards

An emphasis on reducing transactions costs also pervades the second important feature of diffusion-oriented countries, namely, the system of industrial standardization. Of particular importance to the engineering industries, the German system of industrial standardization is unique in the range of intermediate goods and components it covers, the volume of detail it specifies (notably in relation to performance), and the legal status of its norms. This system emerged as part of a conscious effort to promote rationalization in decentralized industries.[25] Though it operates as a quasi-public authority, the system is almost entirely funded and administered by industries. Although the budget of the German standards operation (DIN) is 2½ times that of its French counterpart (AFNOR), the share of this budget provided by all levels of government is less than half that in France.[26] To this must be added the considerable investment German industry makes in providing technical support for the standardization process.

The immediate impact of the standardization system is to reduce transactions costs by providing clearly specified interface requirements for products. At the same time, it fulfills a quality certification function, which is especially important for industrial components. But its indirect effects may be even greater.

In particular, the standardization process itself—and notably the preparation of new standards and the ongoing review of existing ones—provides an important forum for the exchange of technical information both within each industry and with its users and suppliers. Though this information is ultimately rendered public in the published specifications, the long lead times involved in drafting standards, and the relatively small share of the total information generated that is contained in the published standard, ensure that the exchange process operates as a local public good. The primary beneficiaries are the firms most actively involved in industry associations. The density of these information flows also ensures that by the time a new standard is announced, German firms are in a position to adopt it. The system of industrial standardization, in other words, functions as a means of placing ongoing pressure on firms to upgrade their products, while providing them with the technical information required to do so.

Cooperative Research and Development

A concern with assisting a decentralized industrial structure to adjust to changing technologies also underlies the third feature of these countries' policies, namely, the role of cooperative R&D.[27] This takes two forms.

The first is close industry-university links, which have traditionally been of particular importance to the chemical industry and remain a dominant characteristic in Germany, Switzerland, and Sweden. Thus, 15 percent of university research in Switzerland is funded by industry—the highest share in the OECD and more than 3 times higher than in the United States, France, or the United Kingdom. The links go well beyond the chemical sector, as the close ties between the EFTZ in Zurich and the Swiss mechanical and electrical engineering industries attest. Similar links can be found in Sweden, notably between the technical universities and the large science-based firms. A specific feature of the German system is the role of the three large nonprofit research organizations in cooperative research. The Fraunhofer Gesellschaft, in particular, has 22 research centers, which have become increasingly involved in providing technical support to small and medium-size firms.

The second major form of cooperation in R&D centers is industrywide cooperative research laboratories. These account for a considerably higher share of total R&D expenditure in Scandinavia than elsewhere. Thus, in Norway even the largest firms have only small in-house research units, and most industrial R&D is contracted out to cooperative laboratories. In Sweden an extensive network of industry or technology-specific laboratories is jointly funded by industrial firms and by the State Board for Technical Development. In addition to ongoing programs aimed at the entire population of an industry, these laboratories carry out contract research for individual firms. Similar arrangements exist in Germany and (though on a smaller scale and with considerably less government funding) for certain industries in Switzerland.

The most immediate impact of the availability of these outside sources of research expertise is probably on the cost-effectiveness of R&D. They permit sharing of costly instrumentation and research facilities and allow firms occasionally to draw on specialists they could not afford to employ full-time. In this sense, their role is similar to that played by the larger U.S. technical consultancies (for example, Arthur D. Little or Battelle Laboratories) in providing support to smaller laboratories.

This role may be secondary over the longer term, however, as it can be argued that the important function of cooperative research is really twofold. The first is *technology transfer*. Universities and cooperative research centers inevitably have a higher ratio of research to development than have the laboratories of small firms. This higher research intensity allows them to generalize, and hence transfer, the results of individual development projects from firm to firm, thus providing a degree of economies of scope to innovation programs across an industry or activity.

The second function of cooperative research is *technology focusing*.

The process of setting research priorities for the system encourages firms to pool their perceptions of major technological threats and opportunities. This in turn feeds back into the internal R&D planning.

However, the effective discharge of these functions requires that firms have a certain degree of in-house R&D capability, which they complement through recourse to external sources. Thus, the evidence for Germany suggests that the most intensive users of contract research are small and medium-size firms with an internal research unit—on average, these firms spend on external (contract) research an amount equivalent to 30 percent of their in-house R&D spending.

The Role of Policy: An Example

It has therefore been a major concern of policymakers, particularly in Germany, to ensure the existence of an in-house R&D capability to complement other forms of R&D. The Federal Ministry of Economics has in recent years helped finance a scheme providing a partial subsidy for the employment of research scientists and engineers in small and medium-size firms. Assessments suggest that the program has been a considerable success and that about 10 percent of the eligible firms participate. The scheme is worth examining because it provides a particularly good example of German diffusion-oriented policies and notably of what are referred to as indirect specific programs. The latter are government programs specific to the technology of a particular industry but implemented through a trade or industry association rather than by a government department. Three features of these programs stand out.[28]

The first is that the funds involved are small. In total, in 1985 expenditure on the R&D employment subsidy was around 420 million DM—less than 1 percent of German expenditure on R&D. Moreover, the funds were thinly spread, going to about 7,000 firms, a third of which have fewer than 50 employees.

The second is the decentralized process of implementation. The major responsibility for administering the project lies not with the funding agency, but with the German Federation of Industrial Research Associations (AIF), which groups some 90 nonprofit industrial R&D associations, which in turn represent 25,000 firms in 32 industrial sectors. The AIF—70 percent of whose funds come from industry—operates some 60 research laboratories, employing 4,000 scientists and engineers.

Though the AIF has operating responsibility for the project, a low level of discretionary decision making is involved. Eligibility criteria are clearly set out, and question of whether a firm is eligible is straightforward. The risks of discrimination against particular firms are therefore low. However, being administered by the AIF provides the scheme with high visibility among industrial associations, and more than 50 percent of the firms participating in the scheme learned about it from trade associations or local Chambers of Industry and Commerce.

Decentralized implementation is closely related to the third feature of the scheme, namely, the simplicity of its administrative formalities. The application forms do not call for any particular expertise—90 percent of participants completed these forms without any external assistance. This limits the fixed costs involved in participating and further reduces the risks that the program will degenerate into a privileged club.

Defense Research and Development

The importance accorded to the diffusion of technological skills has even affected these countries' not insignificant activities in armaments. Sweden has placed great emphasis on promoting and to some extent organizing the diffusion of defense-related technological skills into the commercial sector. By law, no Swedish company may have more than 25 percent of its business in defense. Thus, defense contractors are forced to develop civilian operations (Gansler, 1980, pp. 245-257). Specific policies have also been implemented to increase the technical capabilities of subcontractors to the larger companies involved in defense work. Financing is provided by the Swedish Industrial Development Fund.

The Effectiveness of the System

The diffusion-oriented countries are therefore characterized by policies that encourage widespread access to technical expertise and reduce the costs that small and medium-size firms face in adjusting to change. In essence, the policy framework serves to increase the capacity for absorbing incremental change without threatening the basic structure of industry.

From this point of view, the policies have indeed been successful. It remains a striking feature of these countries that industrial production is more decentralized than it is elsewhere, notably in mechanical engineering; and that, while providing the benefits of highly focused management, decentralization does not prevent coordination of interdependent decisions and the reaping of economies of scale and scope. Though firms in these countries are smaller than their competitors overseas, higher levels of specialization minimize any relative cost disadvantage.[29]

The system has also functioned effectively in promoting adjustment to incremental change. New skills are transmitted relatively rapidly through labor training and retraining, as well as by interfirm labor mobility. The standardization system itself provides an ongoing flow of technical information; and industry associations and cooperative research institutes allow for interfirm economies of scale in R&D while focusing firms' attention on emerging technologies.

However, two major concerns have been expressed. First, the system as it has evolved is geared to the existing industries, which basically set the technology agenda: That is, they determine the direction of research, dominate the process of standardization, and have a large role in training and education policies. Entirely new industries and technologies may find it difficult to capture the attention they deserve. Second, even in the existing industries, the decentralized, "bottom-up," approach leads to a strong emphasis on movement *along* technological trajectories, while reducing the visibility of, and preparedness for, major shifts in trajectories.

These features—concentration on established industries and moving along set technological trajectories—are apparent in the evolution of these countries' external trade, which has been distinguished by three trends.[30]

The first is that the diffusion-oriented countries have tended to consolidate and even sharpen their traditional patterns of specialization. They have indeed retrenched in the areas where their original performance was poor but without moving into entirely new areas of activity. Rather, their performance has remained strong in the areas where they have traditionally specialized, and *within* these product areas they have tended to become stronger across the board. As a result, their net exports are highly concentrated in "product clusters," mainly in products for which world demand is growing relatively slowly, so that improved performance has required a long-term gain in market share.

Second, this gain in market share has occurred in products with unit values well above the average for their product category. For engineering products, around 85 percent of Swiss exports, 75 percent of German exports, and 65 percent of Swedish exports in 1970 had unit values above the average for their disaggregated product category; this compared with around 35 percent for France and the United Kingdom. Specialization in the higher-quality segments of markets has tended to increase over time.

Third, and most recent, this pattern of specialization has been seriously threatened by competition from Japanese firms, which have used electronics-based technologies to challenge the European countries' traditional predominance in mechanical engineering. Lags in adjusting to shifts in technological trajectories have led to major losses of market share.

These lags arise less from a lack of technological capabilities than from the conservatism inherent in industrywide decision-making processes. The Swiss watch industry and the German machine-tool industry provide striking examples in this respect.

In both cases, the research community associated with the industry was aware of the impact electronics would have—and, in fact, made important contributions to the technology. But research awareness could not be translated into industrial action—partly because of complacency among firms, but also because there were few prospects for adjusting without drastic changes in the industry structure. These changes could not be fitted into the consensus-centered decision-making process; both industries severely lost market share to their Japanese competitors.

Once the loss in market share had begun to occur, however, the industries were relatively well placed to respond. The basic technological skills had been accumulated, and the mechanisms for transferring them to industry were in place. Particularly the German machine-tool industry—which benefited in the early 1980s from the effective devaluation of the DM relative to both the U.S. dollar and the yen—succeeded in reversing its loss of market share and making a quick though painful transition to the new technology.

The criticism that the system slows adjustment to entirely new opportunities while reinforcing specialization in the traditional areas of activity may therefore have some foundation. As these cases bear out, however, the system's capabilities for adjustment—albeit delayed—should not be underestimated. An ongoing response to the Japanese challenge will require important changes in certain aspects of the institutional context. Thus, it has been argued that the apprenticeship system should provide a broader range of generic skills, which could be complemented through continuing vocational education. The Swedish educational reform, which has somewhat reduced the vocational component of secondary education, clearly goes in this direction (Hodenheimer, 1978). But given these changes, the diffusion-oriented countries should remain important on the world industrial scene.

JAPAN

In this typology, Japan is in a class of its own. Like the countries in the first group, it has deployed coordinated efforts to advance national technological goals. At the same time, like the countries in the second group (and with a clear element of imitation from those countries), it has emphasized a broadly based capacity to diffuse innovation-related public goods. In both cases, however, the specific policies and their implementation have been modified to the requirements of the Japanese context.

Two features of this context stand out. The first is that even in the recent past, Japan was at a far lower level of development than the other countries examined in this chapter (Nakamura, 1981; Shinshara, 1970). As late as

1965, Japanese GDP per capita was half the OECD average and less than one-third that in the United States. The gap in per capita GDP was closely linked to lower levels of capital and skill per unit of output throughout Japanese industry. This was accentuated by a dualistic industry structure that combined relatively high productivity in the large-firm sector with considerably lower productivity in the smaller manufacturing firms, in agriculture, and in services. More so than for the other countries in our sample, material well-being in Japan depended on whether Japanese industry could fundamentally reshape its comparative advantage in international trade, rather than simply adapt it to incremental advances in the technological base.

The second factor that sets Japan apart is the relation of the state to industry (Johnson, 1982). Unlike the diffusion-oriented countries, Japan entered the 1950s with an economic bureaucracy able and willing to deploy an active strategy of industrial transformation. Compared to previous periods in its history, and even to the present day, this bureaucracy was at that time uniquely powerful relative to the other political actors on the national scene. However, particularly with the end of postwar reconstruction, the bureaucracy's power depended on its capacity to generate a consensus among the major actors, so that the "administrative guidance" it provided would be smoothly carried through into corporate decision making. This resulted in a combination of consensus-based but relatively centralized decision making with a more decentralized approach to implementation.

The Development Strategy

Combined, these factors have led the Japanese bureaucracy toward a development strategy that emphasizes the rapid upgrading and transformation of the nation's technological skills but does so in a manner both more decentralized and more broadly based than in the mission-oriented countries. The constraints arising from industrial dualism have led to a greater emphasis on diffusion, whereas those arising from the nature of governance have led to a greater emphasis on indirect implementation.

There are three major elements to this strategy: investment in human capital, promoting activities at the "leading edge" relative to the core sector's technological capabilities, and facilitating the transfer of new technologies from the core to the periphery.

Human Capital

A key component of the strategy has been the progressive upgrading of Japan's base of human capital. Japan has a long tradition of engineering training, having been among the first countries to integrate engineering into university curricula (Nakayama, 1984). But its emergence as one of the world's leading centers (at least in numerical terms) for the training of engineers is nonetheless spectacular (Table 6). This has been paralleled by a sustained increase in the average educational attainment of successive cohorts.

The emphasis on the upgrading of human capital is, in many respects, reminiscent of German, Swedish, or Swiss industrialization. But, in contrast to these countries, the expansion of the Japanese skill base has occurred on a more general and less industry-specific basis (Stevens, 1986). More particularly, the Japanese education system is one of general rather than vocational education. The growth in enrollments has consisted in increasing the share of the cohort remaining in the general stream, gradually bringing this share toward U.S. levels (Table 7). Even in postsecondary education, the level of specialization is low, and engineering training in Japan is considerably more superficial than that in Scandinavia or the German-speaking countries.

As a result, the tasks of directing the labor force toward specific occupations and developing the relevant skills has largely been left to industry, and particularly to the larger, "lifetime employment" firms.[31] Firms have had access to a progressively better-educated flow of labor force entrants, notably as regards general mathematical and engineering skills, but little attempt has been made in the education system to shape the capacities of students toward particular vocations. This has given the Japanese labor force a high degree of malleability, decentralizing a set of decisions that critically affect a country's technological capability.

Sectoral Promotion

A high degree of decentralization has also characterized the promotion of particular industries.[32] Three features are important in this respect. First, the areas being promoted have generally been fairly loosely defined, cov-

TABLE 6   Higher Education Engineering Qualifications

Country (Year)	First-Degree Level	Per Million Population	Below First-Degree Level	Per Million Population
Federal Republic of Germany (1981)	7,000	110	16,000	260
United States (1982)	80,000	350	—	—
Japan (1982)	74,000	630	18,000	150

SOURCE: National Economic Development Office (U.K.) and Manpower Services Commission (U.K.).

TABLE 7   Distribution of Students in Upper Secondary Education (Full-time and Part-time Enrollments), Around 1980–1982

Country	General Education (percent)	Vocational and Technical (percent)
Japan	70	30
United States	76	24
Federal Republic of Germany	21	79
France	40	60
Italy	34	65
Netherlands	40	60
United Kingdom	57	43
Switzerland	25	75
Austria	17	83
Belgium	44	56
Denmark	37	63
Finland	50	50
Sweden	30	70

SOURCE: Organization for Economic Cooperation and Development.

ering a broad range of market segments rather than focusing on a particular product. Second, the policies adopted have primarily provided a *framework* within which the activity could develop, notably through import protection, restrictions on foreign direct investment, assistance in licensing overseas technology, and measures aimed at reducing the entry barriers to domestic firms. Beyond providing this framework, policies have rarely involved promotion of a particular domestic company to the apparent detriment of others. Little use has been made of "national champions," and efforts have consistently been made to diversify risk by promoting competition in the domestic market. Third, direct financial assistance has played a very limited role. Though "soft loans" have at times been important, the primary emphasis has been on nondiscretionary instruments such as tax expenditures. Whereas the volume of these tax expenditures may in certain cases have been large relative to the size of the activity being promoted, the subsidies involved have probably been small, partly because of the generally low incidence of corporate taxation. There has, of course, been a subsidy element in public procurement, but given its low defense expenditure, the Japanese government has been a relatively marginal consumer of high-technology equipment (though this is not true in some of the areas discussed below, notably telecommunications and aerospace).

Technology Transfer

The technology transfer policy itself is highly decentralized in Japan, both in implementation and funding (Ergas, 1984b, p. 22). The core of this policy is the network of prefectural laboratories whose primary function is to provide technical assistance in developing *or* adapting new technologies, notably to small and medium-size firms. Central government finances half the laboratories' capital equipment costs, and regional authorities and firms themselves provide the rest of the laboratories' income.

There are now 195 regional laboratories in Japan: The 47 prefectures average four laboratories each, and each prefecture has at least one. Some laboratories are "problem-oriented" (for example, around textiles, food, ceramics, paper, leather, metals) and are linked to the various regional industries and located accordingly. The others (about 30 percent) are more broadly based and multidisciplinary. Generally, they operate in three or four complementary fields (e.g., mechanical engineering, metals, woodworking). The 195 laboratories employ more than 5,000 research technicians and engineers. They are connected with central government laboratories, which provide high-level expertise and sophisticated equipment for R&D when needed. Moreover, the staff of the prefectural laboratories are systematically retrained by the central government to keep them abreast of the latest developments in science and technology. However, the laboratories' activities are determined mainly by their local clients.

Effectiveness

Given the degree of decentralization, notably in the implementation stage, the overall effectiveness of the system has mainly been due to industry's strong response to the opportunities signaled. In part, this response has reflected the high legitimacy the economic bureaucracy enjoys in Japan, so that the advice it provides is taken considerably more seriously by industry than is similar advice in other countries. This legitimacy is reinforced by the fact that a consensus of views with industry is reached well before policies are announced. However, the strength of Japanese industry's response has also been due to a set of factors that have increased the benefits and reduced the costs of exploiting new opportunities.

The first, most obvious, and in some respects most pervasive of these factors is the favorable macroeconomic context. An economy where savings and investment are abundant and where consumer demand is rapidly increasing and shifting to progressively higher-quality goods provides a supportive framework for technological upgrading.

This macroeconomic environment reinforces a second factor contributing to rapid adjustment, notably the low levels of social resistance to change. In addition to steady growth in employment, resistance to change has also been weakened by the lack of strong industry lobbying for declining manufacturing sectors, by the assurances of retraining provided within the lifetime employment system of large firms, and by the "sunset" policies adopted by the Ministry of International Trade and Industry (MITI) to ease the process of decline in industries—such as textiles, shipbuilding, or most recently, aluminum—that have lost their competitiveness (Launer and Ochel, 1985). These factors have also made firms less reluctant to enter new areas, since they know they will be able to withdraw if the opportunities prove ephemeral.

The Role of Competition

A final factor accelerating the response to new opportunities is intense rivalry between the large industrial groups. This rivalry—reflected in far-reaching price competition, in investment "races," and in competition in R&D—is accentuated by several features of the Japanese industrial environment.[33]

The rapid growth of demand—and the perception that growth will continue—has made oligopolistic coordination difficult, while focusing firms' attention on long-term market share rather than short-term profitability. The low cost of funds has reinforced the tendency to take a long view in investment decisions, notably by reducing the implicit discount rate for capital budgeting decisions.[34]

The strategy and structure of Japanese industry also tend to increase the importance of first-mover advantages, so that once a new area emerges, competition to be an early participant is intense (Kono, 1984). Although first-mover advantages in the United States are probably concentrated in the mass marketing stage, production cost factors appear to be more important in Japan. Operating with a fairly fine division of labor relative to smaller enterprises, the major Japanese firms specialize in large-scale fabrication and in mass assembly. These operations are characterized by substantial static and especially dynanic economies of scale. As a result, a large firm's unit costs are highly sensitive both to the rated capacity of its plant and to accumulated production. Given these characteristics, and especially in a rapidly growing market, the penalties of late entry are likely to exceed the costs of building ahead of demand. Market entry and capacity expansion therefore tend to occur quickly as firms seek footholds in new areas of activity.[35]

These pressures are accelerated by each major firm's reliance on a reasonably stable group of smaller suppliers (Imai and Itami, 1981). Unlike the situation in the United States, a late entrant in Japan cannot reduce its costs by acquiring a firm already established and experienced in the business. Moreover, its competitive disadvantage can be aggravated by the fact that its more or less fixed circle of suppliers will also lack experience in the new area. Entering a new market early provides an insurance that the firm will not be severely handicapped should the market prove particularly promising.

Lifetime Employment

The lifetime employment system itself creates strong pressures for large firms to enter emerging markets. Firms committed to lifetime employment seek to diversify their portfolio of activities to cover different stages of the product life cycle, so as to stabilize employment requirements over time. The search for new areas of activities is likely to be a particularly high priority for younger professional staff, given the impact it will have on their career prospects.

The lack of interfirm mobility of managerial staff (who constitute the bulk of the personnel covered by the lifetime employment system),[36] and the consequent need to ensure a sufficient growth to keep internal planning resources fully employed, creates insistent pressures for diversification. However, since diversification must rely on internal expertise, it is largely confined to areas related or similar to the firm's principal activity. Japanese firms consequently tend to expand through related diversification, and the growth of conglomerates is extremely rare (Imai et al., 1984; Nonaka et al., 1983; Kono, 1984).

This creates a system that breeds on itself. The drive for related diversification pushes firms' R&D efforts into adjacent areas. The fact that one firm is seen to do this propels other firms to do the same. Particularly when the technologies involved are generic, in the sense of spanning several product fields, the degree of interfirm competition for footholds in emerging product areas rapidly becomes intense (Suzuki, 1985). This increases the extent of experimentation in the Japanese market, providing a competitive advantage to the economy as a whole.

Interfirm Cooperation

The system involves a high degree of horizontal and vertical cooperation, mainly within each family of firms. The dual structure of Japanese industry is of obvious relevance in this respect. Three factors perpetuate this structure: the problems inherent in a system without well-developed equity markets; the high rigidities associated with internalizing activities into the larger firms, given the lifetime employment system; and an abundant supply of entrepreneurs.

However, this structure could hardly survive without constant upgrading of technological capabilities in the secondary sector. This need is mainly met by the direct technical assistance large firms supply to their smaller subcontractors (Gönenç 1984; Gönenç and Lecler, 1982), but the decentralized laboratory system discussed above also plays an important role, as do trade associations and the standardization system, both originally modeled on German lines.

Another important type of cooperation among firms occurs in the context of precompetitive research, notably for the development of generic technologies. These research efforts, in particular those promoted by MITI, provide for investigatory research in the phases of R&D generating the highest content of information in the "public good." They are to some extent a substitute for university-industry links, which appear to be extremely weak in Japan. Whether they are effective in this respect is the subject of considerable controversy, but cooperative research does appear to have allowed Japanese firms to resolve some of the critical bottlenecks confronting them in the "generic technology" aspects of the industry they were entering: for example, the production of cathode ray tubes for color television receivers (Dore, 1983; Peck and Wilson, 1982).

Centralized Programs

These factors go a considerable way toward explaining the responsiveness of Japanese firms to the signals emerging from Japan's largely decentralized system of industrial planning. However, it would be foolish to deny that the Japanese bureaucracy has at times engaged in highly directive and centralized attempts to promote particular activities and that these attempts have involved a considerable mobilization of resources.

Prominent examples are mainframe computers, central office telecommunications equipment, aerospace, and the Japanese railway plant. Policy in these areas has been similar to that in Europe, with the important difference that a greater number of competing firms have been present in each area. Despite this difference in policy design, the outcomes do not suggest a high level of policy effectiveness: Japan's output of videotape recorders far exceeds its production of computers; Japanese central office electronic switching systems have not emerged as major competitors on world markets; the Japanese bullet train is far less cost-effective than its French counterpart, the Train à Grande Vitesse; and aerospace remains a weak point in the Japanese industrial structure.

Overall Impact

Japan's policies have tended to be most successful when they combine three features: consensus decision making on broad goals, decentralized implementation, and a reliance on the dynamics of competition to ensure a rapid response. Sustained by good macroeconomic management, a steady increase in human capital, and a willingness to adjust to change, this has proved to be a formidable engine of growth.

In particular, it has allowed Japanese firms to modify their specialization in international trade into progressively more technologically advanced product areas (Boltho, 1975; Orléan, 1986). The striking feature of this shift is not only its breadth but its depth: Like the diffusion-oriented countries, Japan's export pattern is highly specialized, the bulk of net exports being concentrated on a small number of comodities. Unlike the diffusion-oriented countries, however, this pattern has shifted markedly over time, as the first wave of Japanese exports (mainly textiles) was replaced by a second (steel, shipbuilding), a third (automobiles), and now a fourth (electronics and machinery). Such drastic changes in international specialization have inevitably entailed major modifications in the structure of Japanese industry. The capacity of the Japanese industrial structure to carry out such shifts is what sets it apart from the other countries considered in this chapter.

However, this does not imply that what has proved successful until now will remain so. Particular concern has been expressed in Japan about whether the system for promoting innovation is resilient.[37] The central question in this respect is the system's continuing effectiveness once Japan arrives at the technological frontier. It is presumably easier to set broad goals in the catching-up stage of growth than in pushing beyond the state of the art. Moreover, the skills needed to implement these goals differ. Up to now, Japan has not suffered from the weakness of its scientific base, but it may prove vulnerable to a blurring of the boundaries between pure and applied research.

SHIFTING AND DEEPENING: AN ATTEMPT AT SYNTHESIS

Directions for Research

In recent years, economists have made significant progress in analyzing technological advance as an evolutionary process—that is, a process of experimentation, selection, and diffusion.[38] The work done provides a convenient analytical structure for synthesizing the arguments presented above and for examining their implications for overall economic performance.

A central concern of recent analyses has been the mechanisms by which innovation shapes market structure, notably its impact on concentration and on the extent of the barriers to potential competition. The assumption has been that this relation operates similarly from country to country; but the data presented above suggest that this is not the case. Rather the material reviewed suggests important differences between countries along three dimensions:

- Who *appropriates* the gains from technological advantage? Is it the innovating firm alone or is it the firm and a broader group (for example, a firm's suppliers)?
- To what extent are these gains *cumulative and sustainable* over time? Where does the process of skill accumulation occur—in the individual firm, in the industry, or in the industrial structure as a whole?
- How much *flexibility* is there in responding to innovation? Does flexibility occur through adjustment by existing firms or through shifts in the firm population?

The material reviewed also suggests that differences in each of these respects affect the evolution of each country's industrial structure. In essence, this relation operates through the balance between two (not necessarily alternative) ways of increasing the efficiency with which resources are used: *shifting*, or transferring resources from old to new uses; and *deepening*, or improving their productivity in existing uses.

The greater the mobility of technical, managerial, and financial resources, the greater the contribution that shifting is likely to make to overall growth. Conversely, the greater the extent to which assets are firm- or industry-specific, the greater the importance of deepening to long-term competitiveness. This relation can be highlighted by reexamining four of the countries in our sample; these countries (the United States, France, Germany, and Japan) can be considered to be roughly representative, given the similarities between the United Kingdom and France, and between Switzerland, Sweden, and Germany. A broad characterization of the four countries is given in Table 8, which summarizes many elements of the discussion above and can be analyzed as follows.

The United States can be considered paradigmatic of shifting. An extremely large applied research system, operating at the frontiers of technology, continuously generates potential new areas of commercial activity. Adjustment to these opportunities occurs through competition between firms on the open market for mobile technical and managerial skills and financial assets. The ease with which these resources can be bid out of existing uses discourages productivity-enhancing investments in skills and capabilities that are specific to a particular firm or activity, as such investment can be justified only through longer-term commitments. However, high mobility also ensures that entirely new areas of endeavor are rapidly exploited, first in the domestic market and then through world sales.

TABLE 8 Technology Systems and Industrial Structures

Characteristic of System	United States	France	Germany	Japan
Appropriation	Firm	State	Firm and industry	Industrial group
Skill accumulation	Labor market	Technocracy	Industry and research system	Large firm
Flexibility	Mainly by entry and exit	Determined through the political system	Adaptation to incremental change; low intersectoral flexibility	High, but major actors remain the same
Industrial structure and trade pattern	Product cycle	Dualism	Inherited specialization	"Moving clusters"

In France, the transfer of resources to new activities does occur, but largely (though not solely) through major state-initiated programs aimed at both public and private markets. The technical elite, which is a more or less integral part of the state apparatus, is the essential repository of technological skills and plays the key role in designing and implementing programs. However, the concentration of power in this elite and the limited diffusion of skills and capabilities outside its area of activity has two consequences. First, the "shifting" is constrained to those parts of the economy directly affected by the large public programs. Second, the rest of the economy lacks the resources (and often the incentives) to "deepen" its competitive advantage.

Germany, in contrast, is paradigmatic of deepening. Skills and resources are highly industry-specific, and their development follows paths largely charted by the industries themselves. Relations between firms, between firms and their employees, and between firms and the financial system have traditionally included long-term commitments favoring investments in activity-specific capabilities. At the same time, high levels of education, industrial standardization, and cooperative research provide powerful mechanisms for diffusing capabilities throughout each industry, so that progress is made across a broad front. The pattern of industrial capabilities is largely inherited, yet it is constantly renewed by "doing what one has always done, but better."

The distinctive feature of Japan is the extent to which it combines shifting with deepening. The key component is the large firm, closely linked to its main sources of finance and surrounded by a network of smaller suppliers. The large firm—and more generally the industrial group—seeks to maximize the productivity with which resources are employed in existing uses. However, it also faces powerful incentives to shift its operations toward emerging areas of activity, bringing the entire industrial structure in its wake. Three factors are at work: first, the long-term nature of commitments permits productivity-enhancing investments in firm-specific skills; second, the intensity of competition between large firms encourages early entry into new markets; third, each large firm, as it moves, seeks to shift its suppliers with it.

Implications for Overall Economic Performance

But between what does one shift, and along what does one deepen? And what implications does the balance of shifting and deepening have for overall economic performance?

The concept of a technological trajectory provides a helpful building block in exploring these questions. A technological trajectory can be defined as a path of technological development, drawing on a given set of basic scientific principles and propelled by an internal dynamic of improving performance with regard to a few key design criteria (Dosi, 1982; Nelson and Winter, 1982; Rosenberg, 1976). At the risk of considerable simplification, evolution along this path can be characterized as following an S-shaped curve (Figure 1):

- The emergence phase includes experimentation among alternative design approaches, as attempts are made to identify approaches with the greatest promise for subsequent developments.
- In the consolidation phase, the concentration of R&D on a few critical parameters, within the framework of a broadly set design approach, allows rapid improvement both in performance and in cost.
- The maturity phase occurs as the most easily exploited opportunities have been fully used, while entirely new design approaches, possibly based on an area of applied science different from that of the original trajectory, emerge as substitutes in a growing range of uses.

The development of vacuum tube technology illustrates these processes and their pattern of evolution over time (Baker, 1971; Maclaurin, 1949; Sturmes, 1958). After a phase of open experimentation, Lee de Forest's triode tube set an underlying structure for the workable amplification of small electrical signal voltages. Subsequent progress in tube technology, though yielding dramatic improvements in performance, concentrated on a few variables, such as the energy efficiency of the cathode, tube life and reliability, and automation of the manufacturing process. However, the development of solid-state semiconductor technology beginning in the

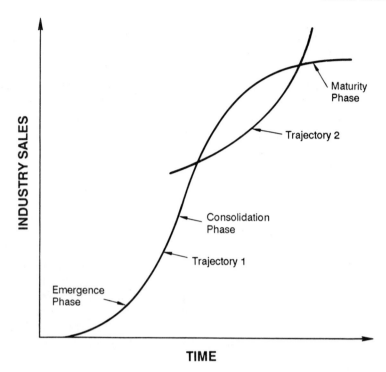

FIGURE 1   Technological trajectories.

late 1940s dramatically cut across this path of improvement (Webbink, 1977). Transistor-board devices rapidly established themselves as a more reliable and space-saving alternative to the vacuum tube, with enormous potential for cost reduction through progressively larger-scale integration and automated manufacturing and testing.

As the technology developed, so the structure of the industry changed. In the early days of the vacuum tube industry, the field was open to competition. With many differing approaches to tube design, manufacturing, and application, overall profitability in the industry was probably low, since the small number of "hits" was more than offset by high initial development costs and a large number of "misses" (de Forest himself suffering repeated bankruptcies). Profitability increased only after the basic technology had stabilized, and patents and proprietary know-how blockaded entry, weakening price competition, improving R&D focus, and allowing cost reduction as output grew. The industry's consolidation phase was dominated by a tight-knit oligopoly, including some of the largest, most technologically advanced firms of its day: General Electric, Westinghouse, RCA, and AT&T in the United States; Marconi, Siemens, and Philips in Europe.

Large size and (for the time) huge R&D budgets did not allow these firms to transfer their dominance to the emerging market for solid-state devices. These drew on an applied science base quite different from that they had mastered over the years. However, the vacuum tube industry did not disappear, for four reasons: initial uncertainty about the capabilities of solid-state devices slowed substitution, the emergence of solid-state competition encouraged manufacturers to bring forward improvements in tubes, rapid growth occurred in applications where there were no practicable substitutes for tubes (notably television receivers), and new tubes were developed for applications requiring frequencies unsuitable to solid-state technologies. Substantial opportunities persisted in the industry 40 years after its technological base had been superseded, but these opportunities relied on a progressively narrower and more vulnerable base.

Three broad conclusions can be drawn from this account:

- The emergence of a technological trajectory is not usually associated with high rates of return on investment, given large R&D costs, the substantial risk of failure, and the intensity of competition.
- It is in the consolidation phase that the greatest improvements are made in product cost and performance and the largest scope exists for supranormal profits.
- As improvements in critical parameters become more difficult to achieve, the maturity phase creates new challenges for the industry, with the development of substitute products intensifying competition and increasing the importance of capturing the least vulnerable niches.

Clearly these conclusions do not have the force of laws, nor can one indiscriminately generalize from the level of individual industries to that of national industrial structures.[39] Nonetheless, they suggest several hypotheses of interest:

- The performance of an industrial structure specializing in the emerging phase is likely to depend first on its capacity to experiment on a broad front, thus increasing the probability of success. An important factor in this respect is proximity to a pool of sophisticated customers, who can rapidly distinguish promising from less promising alternatives. Second, performance will depend on the extent to which the industrial structure can carry successes over from the emergence to the consolidation phase. However, there is no *a priori* reason to expect such an industrial structure to show a high rate of growth of real incomes or productivity, at least as conventionally measured (Ergas, 1979).
- Conversely, an industrial structure specializing in the consolidation

phase can expect to capture substantial gains in productivity and per capita income. Whether these gains will persist, however, depends on the capacity of the industrial structure (a) to exploit the results of successive emergence phases without having fully borne their costs, and (b) to transfer resources from one technological trajectory to another as the maturity phase sets in.

- To succeed, an industrial structure pursuing technological trajectories into the maturity stage will require high levels of efficiency both in R&D and in applications engineering, allowing it (a) to obtain a maximum of performance improvements out of a given path of development, thus slowing the substitution process, and (b) to retain profitability by specializing in the product segments least vulnerable to intensified competition. Nonetheless, it may be supposed that the long-term performance of such an industrial structure will be constrained by the gradual slowing of market growth and the decreasing number of technological opportunities.

These hypotheses merge naturally with the country analysis presented below.

Thus, the predominance of "shifting" behavior in the U.S. economy corresponds to specialization in the emergence phase of technological trajectories. The returns to this pattern of specialization are maximized by (a) the scale on which experimentation occurs, increasing the probability of success; (b) the sophistication of the U.S. market (including its public procurement component), which accelerates the process of selection among competing alternatives; (c) the rapidity with which breakthroughs in the noncommercial parts of the technological system diffuse into the sphere of commercial experimentation; and (d) the existence of a substantial pool of large U.S. firms capable of transferring the results of experimentation in the U.S. market into world sales.

However, the inherent characteristics of this phase of technological evolution limit the rate of growth of per capita incomes to which it can give rise. These limits have been accentuated by the declining competitiveness of U.S. sites (though less so of U.S. firms) in the mass production operations characteristic of the consolidation phase.

The "imperfect shifting" that is sometimes considered a major feature of France's technological system limits the returns obtained from concentrating on the emergence stage of technological trajectories. High levels of investment in R&D are incurred to establish a presence in this stage, although the scale of experimentation may still be too small to achieve a reasonable chance of success across the board. Even when successful outcomes are obtained, numerous factors slow their transfer from the mission-oriented environment to that of commercial exploitation and hence the prospects for going from emergence to consolidation.

The growth of French incomes has therefore depended heavily on sectors such as motor vehicles, tires, and food processing, which are outside of—and only weakly linked to—the core technological system. However, performance in these sectors has proved difficult to sustain. This is partly because the decline of traditional industries and the implicit protection accorded high-technology activities has forced other sectors to bear a disproportionate share of unfavorable macroeconomic developments.

At the other extreme, the deepening processes characteristic of Germany's industrial structure are associated with far-reaching specialization in pursuing technological trajectories into their mature phases. An institutional framework that is, in many respects, uniquely suited to this pattern has allowed German industry to exploit fully the higher value-added segments of the markets in which it operates. The experience of the last decades, however, has highlighted some of the risks this pattern of specialization entails. In particular, it creates vulnerability on two fronts:

- Up-market, from competitors operating in the same product markets, but exploiting new technological trajectories as they enter the consolidation phase. These competitors are well placed to provide rapid rates of increase in cost-to-performance ratios—as Japanese firms have done in numerically controlled machine tools; and
- Down-market, from competitors whose technological capabilities may lag, but whose factor costs are substantially lower.

The slowing of total factor productivity growth as technological opportunities along the original trajectory diminish, combined with the rationalization pressures arising from greater rivalry on world markets, could make rising living standards more difficult to achieve. This could endanger the high degree of social consensus that underpins the diffusion-oriented countries' industrial model.

Between these extremes of shifting among emerging trajectories and deepening along mature trajectories, Japan has been extraordinarily successful in exploiting successive trajectories in their consolidation phase. Concentration on this phase has provided numerous advantages to Japanese firms:

- By avoiding the stage of greatest technological and commercial uncertainty, the return on scarce R&D capabilities could be maximized.
- Entering activities at the consolidation (rather than emergence) stage also minimized the importance of close proximity to sophisticated users—until recently a major constraint on Japanese competitiveness.

- Greatest benefit could be drawn from accumulated skills in managing large-scale fabrication and assembly processes and from the cost-reducing pressures of competition for a growing market.

Rapid growth of real income has been achieved by exploiting these advantages. At the same time, Japanese firms' share of world markets has increased at the expense of firms more specialized in the emergence or maturity stages of technological development. This pattern of specialization has relied on a high degree of social consensus, which has made it possible to shift resources rapidly from one trajectory to another. It has also relied on:

- an elaborate network for gathering information from abroad about emerging technologies;
- the low-cost availability of the results of U.S. R&D on entirely new technologies; and
- access to world markets in which to achieve economies of scale.

These premises now appear vulnerable in several respects. Access to U.S. technology is not as easy as it once was, mainly because U.S. firms now consider Japanese firms as major rivals. Moreover, Japanese technological performance has passed the stage at which large improvements could be obtained simply by learning from overseas. Finally, access to world markets is threatened by the spread of protectionist measures. Nonetheless, the capacity of Japanese industry to meet these threats should not be underestimated, Japanese R&D capabilities are now more than sufficient to engage in highly advanced research, the Japanese domestic market is large and sophisticated enough to provide a good seedbed for experimentation, and Japanese firms have established the global brand image and distribution channels needed to sell a more diversified range of products internationally.

This discussion suggests that there are different paths to happiness, as countries' institutional structures and social arrangements facilitate specialization in differing stages of technological evolution (see Figure 2). Each of these stages has advantages and disadvantages in providing for the growth of real income, but countries also differ in the extent to which they succeed in securing the greatest benefits from any given pattern of specialization.

Over the longer term, these differences in R&D efficiency may be most important. Consider France and Germany: The French state has encouraged specialization in the emergence phase of technologies, whereas German industry has largely retained its traditional pattern of specialization. However, the disparities in performance among these countries arise less from this difference in specialization than from the efficiency with which the

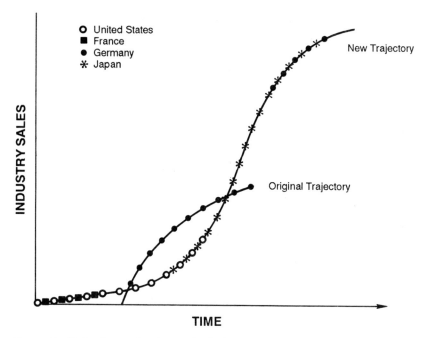

FIGURE 2   National strengths along technological trajectories.

potential economic gains implicit in each pattern of specialization are exploited. In other words, location on a technological trajectory may be less important than the efficiency with which the advantages of that location are pursued. This, in turn, depends on institutional features (broadly defined) that may be more or less appropriate for a given pattern of specialization.

It is by no means obvious that the institutional features typical of one economy can be transplanted to another. But a few general factors underlie the differing outcomes countries obtain from similar patterns of specialization. It is to these factors and particularly their implications for policy that we now turn.

POLICY IMPLICATIONS

The dominant feature of national technological systems is *diversity*. This partly reflects differences in policy stance between countries, but many other factors are also at work. Examination of these factors suggests several conclusions relating to the scope and limits of technological policy.

The first and most fundamental is the dependence of technology policy outcomes on their economic and institutional environment. The policies pursued in the United Kingdom or France do not differ greatly from those of the United States, but the outcomes do. The reasons for this lie partly in the details of policy design and in the manner in which policies are implemented. But deeper and more pervasive factors are of far greater significance.

In part, the U.S. advantage arises from the very size of its scientific and technological system. This ensures that mission-oriented research crowds out commercial R&D to only a limited extent and that there is a huge stock of firms and individuals capable of absorbing and commercializing the results of mission-oriented research. But this advantage of size is accentuated by other features of the U.S. system.

In particular, new technological capabilities spread rapidly in the U.S. economy, both through the direct transmission of ideas—for example, between industry and university—and through the high mobility of technologically skilled personnel. Moreover, lower entry barriers into U.S. industry, combined with pressures for firms to be among the early entrants into new product markets, accelerate the transformation of technological advances into commercial innovations.

In France, by contrast, several factors slow the transfer of the technological advances generated by mission-oriented research into the commercial sector. These include the paucity of contacts between universities and industries, the low mobility of scientists and engineers, the pervasive obstacles to the entry of new firms, and the protective atmosphere of government procurement in which larger firms prefer to remain.[40] Those differences mean that in the United States the results of government-supported R&D diffuse quickly into the commercial sector of the economy, but in France, and even more so the United Kingdom, they remain more or less confined to their sector of origin.

The Importance of Diffusion

This suggests a second conclusion, which is that the key problem of technology policy (as distinguished from science policy) lies less in generating new ideas than in ensuring that they are effectively used. The "high-technology industries," however defined, are inevitably a small part of total output; taken on its own, even predominance in these industries will have a limited impact on living standards (Nelson, 1984; Riche et al., 1983). Rather, long-term growth mainly depends on the capacity to deploy technical capabilities across a broad range of economic activities.

This goal can be achieved in various ways. In the United States, the diffusion of technology is largely a market-driven process, which relies on high levels of mobility of human and financial resources and the existence of a marketplace of ideas. In Germany and Switzerland, in contrast, organized social mechanisms for promoting technology diffusion play a more important role—these include the apprenticeship system, the system of industrial standardization, and the network of cooperative research.

Seen purely in institutional terms, these experiences are not easily transferable among countries. Japan borrowed heavily from overseas in designing its institutional framework, but at an early stage of industrial development. It is questionable whether policymakers in the United Kingdom or France could quickly set up processes of industrywide technological cooperation akin to those that developed over a long period of time in the German-speaking countries. The institutional mechanisms for technology diffusion must inevitably reflect broader features of a country's economic, social, and even political environment. However, there are common elements to the countries with a record of success in technology diffusion. These elements can provide a useful indication for technology policy. Three such elements emerge from this study.

Investment in Human Capital

The first element in the successful diffusion of technology is the role of *investment in human capital*. This investment has both a flow and a stock dimension. The flow of newly trained personnel into the active population allows the continuous upgrading of skills and capabilities. At the same time, the better educated the labor force is, the greater will be its capacity to adjust to sophisticated new techniques. Higher levels of education are also likely to make this capacity more widespread, both throughout industry and throughout the active population.

Countries whose investment in human capital lacks depth or breadth may be among the pioneers in generating new technologies, given a sufficiently strong scientific elite. But as far as using these technologies is concerned, they will be disadvantaged on two counts: an inadquate rate of expansion or replacement of the skill base at the margin and difficulties in adjusting the existing stock to the demands of technological change. Moreover, their difficulties are likely to persist or even mount. The production of human capital is highly intensive in human capital, and the lags involved in correcting deficiencies in the human capital stock can be extremely long (Sandberg, 1979).

Policy Decentralization

A second factor in promoting diffusion relates to the design of technology policies. Whether those policies actually promote the best use of technological advance appears to be closely related to the range of actors they involve—that is, to their degree of *decentralization*.

This, it can be conjectured, occurs for three reasons. First, centralized programs frequently concentrate resources on the wrong areas. In both the United Kingdom and France, for example, excessive resources have been devoted to projects that are technologically glamorous but not economically relevant. Second, the concentration of resources on a small number of projects itself increases the risk of costly failures, particularly when each project being supported entails a high level of risk. Finally, even if successful in terms of their immediate objectives, large, centralized projects usually pose considerable problems of technology transfer once the R&D phase is completed.

Program decentralization can be achieved in different ways. In the United States, the very scale of the defense R&D program is such that a fairly high level of dispersion of funds is almost inevitable, but conscious policy choices—such as the emphasis on second-sourcing and the support of R&D by new and small firms—are also significant. In Germany, Switzerland, and, to a lesser extent, Sweden, the delegation of policy-setting and implementation functions to industry associations and regional bodies averts the risks inherent in centralized, bureaucratic decision making. The Japanese emphasis on consensus probably plays a similar role.

But abstracting from these differences, similarities emerge. The risks of placing too many eggs in one basket (and choosing the wrong basket at that) can be reduced by making support policy less *discriminatory* in the range of firms and sectors covered and by placing less emphasis on *discretionary* choices among alternative approaches. This implies a preference for measures with a high degree of automaticity—for example, tax expenditures. It also implies preference for the delegation of power and public support to broadly based rather than narrowly based groups—for example, to an industry or research association as a whole rather than a formal "club" of subsidy receivers.

Traditionally, the major argument against nondiscretionary policies is that funds may be provided to firms for projects that would have been carried out in any case. Equally, the case against decentralizing decision making rests on the risk that support programs will be "captured" by organized interest groups who will abuse them to advance narrow sectional concerns. However, experience suggests that the risks of capture are greatest when decisions are highly centralized, since this usually leads to a symbiotic relationship between a small number of policymakers and a few large firms. Experience suggests further that it is in this situation that public support is most likely to become a permanent feature of the cash flow of a narrow range of privileged firms (Bauer and Cohen, 1981; Cawson et al, 1985; Cohen and Bauer, 1985; Young and Lowe, 1974).

Providing Incentives

Even an improved policy framework need not lead to better performance if the *incentives* to make the best use of technological resources are too weak. At a most obvious level, this is a problem of ensuring that firms are exposed to competition so that ideas are quickly transferred from the research environment to that of commercial use.

The problem of providing adequate incentives merits particular attention in three areas: public research laboratories and other nonprofit research institutions, publicly funded commercial R&D, and public procurement. The first of these areas should include scope—notably in the United Kingdom and France—both for reducing the share of public laboratories in government R&D expenditure and for shifting a greater part of their recurrent funding onto a matching grant basis. In the second area, opportunities should be explored for building incentives for success into the system of public support for commercial R&D—for example, by making access to continuing finance more clearly conditional on past performance. The third area, public procurement—notably of complex technological systems—too often serves to subsidize long-term inefficiency rather than to encourage the best use of resources and capabilities. Dismantling these protective devices could impose short-term costs, but these are likely to be small in relation to the longer-term benefits.

In summary, it is true that the institutional framework of any one country cannot be mechanically transplanted to others. Nonetheless, comparative analysis suggests three priority areas for action:

- easing constraints and rigidities that slow the diffusion of new skills and technical capabilities;
- improving the human capital base while enhancing the efficiency of markets for highly trained personnel; and
- increasing the extent to which technology policy relies on market signals and incentives, rather than on the administrative allocation of resources.

ACKNOWLEDGMENTS

The author thanks Bruce Guile of the National Academy of Engineering; Rolf Piekarz of the National Science Foundation; Professor David Encaoua, of the *Direction de la Prevision, Ministère de l'Economie, des Finances, et de la Privatisation*; Christian Sautter, Inspecteur Général des Finance; and P. D. Henderson, H. Fest, J. Shafer, D. Baldes, and many other colleagues at the Organization for Economic Cooperation and Development for their valuable comments on earlier drafts of this paper. Special thanks are also given to the author's colleagues Rauf Gönenç, Andreas Lindner,

Anders Reutersward, and Barrie Stevens for generously providing data and advice. However, the author wishes to stress that unless otherwise indicated, the views expressed in this paper are attributable only to the author in a personal capacity and not to any institution.

NOTES

1. See especially Rosenberg and Birdzell, 1986. Nove, 1983, provides an interesting comparison by the relatively sympathetic description of the functioning of a socialist economy and of its difficulty in innovating.
2. This is a key component of the classic "market failure" argument for public support for R&D. See Antonelli, 1982; Freeman, 1974; Kamien and Schwartz, 1982; Mowery, 1983a; Rothwell and Zegveld, 1981.
3. This description of the United Kingdom draws on Carter, 1981; Dickson, 1983; Hall, 1980; Henderson, 1977; Hogwood and Peters, 1985; Vernon, 1974; Young and Lowe, 1974.
4. This description of France draws on Bauer and Cohen, 1981; Cawson et al., 1985; Cohen and Bauer, 1985; Dupuy and Thoenig, 1983; Grjebine, 1983; Shonfield, 1965; Stoffaes, 1984; Vernon, 1974.
5. See especially Ponssard and Pouvoirville, 1982. The high concentration levels of overall transfers from the state to industry (including public procurement) are discussed in Centre d'Economie Industrielle, n.d., and Commissariat Général du Plan, 1979.
6. On telecommunications see Cohen and Bauer, 1985; Darmon, 1985; Ergas, 1983b; Peterson and Comes, 1985. On energy, see specifically Feigenbaum, 1985; Picard et al., 1985.
7. This discussion of the United States draws on Fox, 1974; Gansler, 1980; Nelson, 1982, 1984; Phillips, 1971; Research & Planning Institute, Inc., 1980.
8. Thus, Scherer (1982) estimates that in the United States only 12 percent of 1974 defense R&D funding generated technologies that flowed directly to clearly nondefense uses.
9. Secondary effects are examined by, among others, Ettlie, 1982; Henderson, 1977; Malerba, 1985; Rothwell and Zegveld, 1981; Scribberas et al., 1978; Teubal and Steinmueller, 1982. An interesting international comparison of secondary effects can be obtained by contrasting U.K. and U.S. surveys of the effects on defense funding on national semiconductor industries: Dickson, 1983; Mowery, 1983b.
10. The role of U.S. government funding in the growth of small firms is discussed in Bollinger et al., 1983; Research & Planning Institute, Inc., 1980. A survey is given in Ergas, 1984b. Defense funding of university research and its growing importance is discussed in National Science Board, 1986, chap. 2.
11. Compare Katz and Phillips, 1982, and Lavington, 1980.
12. See especially Freeman, 1971; Freeman, 1976; National Science Foundation, 1985. Compare with Wilson, 1980.
13. Compare National Science Board, 1986, p. 86 and appendix table 4-17; Le Monde, 6 February 1986.
14. See Ergas, 1984b, pp. 10-11. A fascinating case study is National Academy of Engineering, 1982. The role of scale economics in intensifying rivalry in the transition to mass production is clearly brought out by recent literature on strategic competition. See, for an excellent survey, Kreps and Spence, 1985.
15. See especially Schmalensee, 1982. Advertising-related product differentiation also

appears to be a particularly significant factor explaining persistent profitability in U.S. industry. See Geroski, 1985; Mueller, 1985.

16. Aspects of this pattern are highlighted in Prais, 1981. Robson et al. (1985) examine the diffusion of technology in the United Kingdom. See also the analysis of the United Kingdom's trade structure in Orléan, 1986.

17. See, in addition to the references in note 4 above, analyses of France's trade patterns presented in Lafay, 1985; Orléan, 1986; Vellas, 1981.

18. The results of Lipsey and Kravis, 1985, conflict with those of Dunning and Pearce, 1985, who find a sharper decline in U.S. firms' overall share of revenues and profitability.

19. The classic formulation of this process is Vernon, 1966. For empirical analysis of U.S. trade patterns, see inter alia the contrasting results set out in Hatzichronoglou, 1986; Lafay, 1985; Leamer, 1984; Vernon, 1979.

20. The general characteristics of these countries are explored in Katzenstein, 1985a and 1985b.

21. On Germany and Switzerland, see Henderson, 1975; Milward and Saul, 1977. On Scandinavia, see Hecksher, 1984; Hildebrand, 1978.

22. The general characteristics of these educational systems, and international comparisons, are set out in Stevens, 1986. See also Organization for Economic Cooperation and Development, 1979; Prais and Wagner, 1983a and 1983b; Worswick, 1985.

23. A recent survey reports that in Germany 45 percent of labor force participants with vocational training at a high school level undertook continuing training during the period 1974-1979.

24. According to population census estimates, some 50 percent of the civilian labor force in Germany and Switzerland has completed an apprenticeship. See Organization for Economic Cooperation and Development, 1986.

25. The classic study is Brady, 1934.

26. Estimates are provided in Laboratorio di Politica Industriale, 1982. The literature on standardization is reviewed in Ergas, 1984b.

27. I am indebted to my colleagues in the Science, Technology and Industry Directorate of the Organization for Economic Cooperation and Development for assisting me in compiling the information presented here.

28. See especially Meyer-Krahmer et al., 1983. My colleague Andreas Lindner provided me with particularly useful information on the subjects discussed in this section.

29. See George and Ward, 1975; Prais, 1981; Pratten, 1976. Particularly useful case studies are Aylen, 1982; Daly and Jones, 1980.

30. This discussion draws on Aglietta and Boyer, 1983; Leamer, 1984; Ohlsson, 1980; Orléan, 1986. A particularly useful discussion of the balance between shifting resources among competing uses, as against increasing their productivity in existing uses, is in Carlsson, 1980.

31. It has been estimated that Japanese firms' total expenditure on vocational education is 5 times greater than public expenditure on vocational education.

32. See especially Collins, 1981, 1982; Saxonhouse, 1984; Uena, 1977.

33. See Caves and Uekasa, 1976. On price competition, see Encaoua et al., 1983.

34. A theoretical model in which collusion is less stable in a growing than in a declining market is set out in Rotemberg and Saloner, 1984. Estimates of the cost of funds are given in Ando and Auerbach, 1986.

35. Thus, in the United States, the number of large takeovers (valued at $100 million or

more) has increased steadily over the last decade, rising from 14 in 1975 to 116 in 1982; in Japan, in contrast, the number of large transfers (exceeding $50 million) has been virtually constant, with only 10 such transfers occurring in 1981. See Organization for Economic Cooperation and Development, 1984b.

36. See Tachibanki, 1984, who estimates that lifetime employment applies to no more than 10 percent of the Japanese labor force, almost entirely at higher levels of educational attainment.

37. On the blurring of frontiers between basic and applied research, see Committee on Science, Engineering, and Public Policy, 1983; on its implications for Japan, and concern about the future, see Sciences and Technology Agency (Japan), 1985.

38. Useful overviews are in Antonelli, 1982; Bollinger et al., 1983; Dosi, 1982; Kamien and Schwartz, 1982.

39. Some of the caveats in this respect are set out in Ergas, 1983a. See also Clark, 1985.

40. The fact that France and, to a lesser extent, the United Kingdom have lagged in applying competition policy to their respective national industries has also presumably been a factor reducing the pressure on firms to innovate.

REFERENCES

Aglietta, M., and R. Boyer. 1983. Pôles de Compétitivité, Stratégie Industrielle et Politique Macro-economique. Paris: Working Paper CEPREMAP No. 8223.

Ahlström, G. 1982. Engineers and Industrial Growth. London: Croom Helm.

Ando, A., and A. Auerbach. 1986. The Corporate Cost of Capital in Japan and the U.S.: A Comparison. Research Working Paper No. 1762. Cambridge: National Bureau of Economic Research.

Antonelli, C. 1982. Cambiamento Tecnologico e Teoria dell'Impresa. Torino: Loescher Editore.

Arocena, J. 1983. La Création d'Enterprise. La Documentation Française, 1983.

Aylen, J. 1982. Plant size and efficiency in the steel industry: An international comparison. National Institute Economic Review 100 (May).

Baker, W. J. 1971. A History of the Marconi Company. New York: St. Martin's Press.

Bauer, M., and E. Cohen. 1981. Qui Gouverne les Groupes Industriels? Paris: Editions du Seuil.

Beer, J. J. 1959. The Emergence of the German Dye Industry. Harmondsworth: Penguin.

Ben-David, J. 1968. Fundamental Research and the Universities. Paris: Organization for Economic Cooperation and Development.

Berger, S. D., ed. 1981. Organizing Interests in Western Europe. Cambridge: Cambridge University Press.

Bollinger, L., K. Hope, and J. M. Utterback. 1983. A review of literature and hypotheses on new technology-based firms. Research Policy 12 (February).

Boltho, A. 1975. Japan—An Economic Survey. Oxford: Oxford University Press.

Brady, R. 1934. The Rationalization Movement in German Industry. Berkeley, Calif.: University of California Press.

Brooks, H. 1983. Towards an efficient public policy: Criteria and evidence. In Emerging Technologies, H. Giersch, ed. Tubingen: J.C.B. Mohr (Paul Siebeck).

Carlsson, B. 1980. Technical Change and Productivity in Swedish Industry in the Post-War Period. Stockholm: The Industrial Institute for Economic and Social Research, Research Report No. 8.

Carter, C., ed. 1981. Industrial Policy and Innovation. London: Heinemann.

Caves, R., and M. Uekasa. 1976. Industrial Organization in Japan. Washington, D.C.: Brookings Institution.

Cawson, A., P. Holmes, and A. Stevens. 1985. The Interaction Between Firms and the State in France: The Telecommunications and Consumer Electronics Sectors. Cambridge: Trinity Hall, December 10-13, 1985, mimeo.

Centre d'Economie Industrielle. n.d. Quelques Réflexions à Propos des Mécanismes de Transfert Etat-Industrie Mis en Oeuvre en France et en Allemagne. Centre d'Economie Industrielle, Les Milles, mimeo.

Clark, K. B. 1985. The interaction of design hierarchies and market concepts in technological evolution. Research Policy 14 (October).

Cohen, M., and M. Bauer. 1985. Les Grandes Manoeuvres Industrielles. Paris: Pierre Belfond.

Collins, E. 1981. Corporation Income Tax Treatment of Investment and Innovation Activities in Six Countries. Washington, D.C.: National Science Foundation.

Collins, E., ed. 1982. Tax Policy and Investment in Innovation. Washington, D.C.: National Science Foundation.

Commissariat Général du Plan. 1982. Aides à l'industrie. Mimeo.

Committee on Science, Engineering, and Public Policy. 1983. Frontiers in Science and Technology: A Selected Outlook. New York: W. H. Freeman.

Daly, A., and D. T. Jones. 1980. The machine tool industry in Britain, Germany and the United States. National Institute Economic Review 92 (May).

Darmon, J. 1985. Le Grande Dérangement: La Guerre du Téléphone, France: J.-C. Lattès.

Dickson, K. 1983. The influence of Ministry of Defense funding on semiconductor research and development in the United Kingdom. Research Policy 12 (April).

Dore, R. 1983. A Case Study of Technology Forecasting in Japan—The Next Generation Base Technologies Development Programme. London: The Technical Change Centre.

Dosi, G. 1982. Technological paradigms and technological trajectories. Research Policy 11 (June).

Dunning, J. H., and R. D. Pearce. 1985. The World's Largest Industrial Enterprises, 1962-1983. New York: St. Martins.

Dupuy, F., and J.-C. Thoenig. 1983. Sociologie de l'Administration Française. Paris: Armand Colin.

Earle, E. M. 1986. Adam Smith, Alexander Hamilton, Friedrich List: The economic foundations of military power. In P. Paretz, ed., Makers of Modern Strategy. Princeton: Princeton University Press.

Encaoua, D., with P. Geroski and R. Miller. 1983. Price Dynamics and Industrial Structure: A Theoretical and Econometric Analysis. Paris: Organization for Economic Cooperation and Development (Economic and Statistics Department Working Paper No. 10), July, mimeo.

Ergas, H. 1979. Biases in the Measurement of Real Output Under Conditions of Rapid Technological Change. Organization for Economic Cooperation and Development (Expert Group on the Economic Impact of Information Technologies, Working Party on Information, Computer and Communications Policy).

Ergas, H. 1983a. The Inter-Industry Flow of Technology. Paris: Organization for Economic Cooperation and Development (Workshop on Technological Indicators and the Measurement of Performance in International Trade). September.

Ergas, H. 1983b. Telecommunications Policy in France. Mimeo.

Ergas, H. 1984a. Corporate strategies in transition. In A. Jacquemin, ed., European Industry: Public Policy and Corporate Strategy. Oxford: Oxford University Press.

Ergas, H. 1984b. Why Do Some Countries Innovate More Than Others? Brussels: Centre for European Policy Studies Paper No. 5.

Ettlie, J. E. 1982. The commercialization of federally sponsored technological innovations. Research Policy 11 (June).

Feigenbaum, H. B. 1985. The Politics of Public Enterprise: Oil and the French State. Princeton: Princeton University Press.

Floud, R. 1984. Technical Education 1850-1914: Speculation on Human Capital Formation. London: Centre for Economic Policy Research, April, mimeo.

Forman, P. 1974. Industrial support and political alignments of the German physicists in the Weimar Republic. Minerva (January).

Fox, J. R. 1974. Arming America: How the U.S. Buys Weapons. Cambridge, Mass.: Harvard University Press.

Freeman, C. 1974. The Economics of Industrial Innovation. Harmondsworth: Penguin.

Freeman, R. B. 1971. The Market for College Trained Manpower. Cambridge, Mass.: Harvard University Press.

Freeman, R. B. 1976. The Overeducated American. New York: Academic Press.

Gansler, J. S. 1980. The Defense Industry. Cambridge, Mass.: MIT Press.

George, K. D., and T. S. Ward. 1975. The Structure of Industry in the EEC: An International Comparison. Cambridge: Cambridge University Press.

Geroski, P. A. 1985. Do Dominant Firms Decline? University of Southampton Discussion Paper in Economics and Econometrics No. 8509, August, mimeo.

Glover, I., and P. Lawrence. 1976. Engineering the miracle. New Society 30 September.

Glover, R. W. 1974. Apprenticeship in America: An Assessment. Proceedings of the Industrial Relations Research Association, December.

Gönenç, R. 1984. Electronisation et Réorganisation Verticales dans l'Industrie. Thèse de Troisième Cycle, Université de Paris, Nanterre.

Gönenç, R. 1986. Changing investment structure and capital markets. In H. Ergas, ed., A European Future in High Technology? Brussels: Centre for European Policy Studies.

Gönenç, R., and Y. Lecler. 1982. L'électronisation Industrielle au Japon. Sciences Sociales du Japon Contemporain No. 2. Paris: Centre National de la Recherche Scientifique, Centre du Documentation Sciences Humaines, mimeo.

Grjebine, A. 1983. L'état d'Urgence. Paris: Flammarion.

Hall, P. 1980. Great Planning Disasters. Berkeley, Calif.: University of California Press.

Hatzichronoglou, T. 1986. International trade in high technology products: Europe's competitive position. In H. Ergas, ed., A European Future in High Technology? Brussels: Centre for European Policy Studies.

Hecksher, E. F. 1984. An Economic History of Sweden. Cambridge, Mass.: Harvard University Press.

Henderson, P. D. 1977. Two British errors: Their probable size and some possible lessons. Oxford Economic Papers 29 (July).

Henderson, W. O. 1975. The Rise of German Industrial Power 1834-1914. Berkeley, Calif.: University of California Press.

Hildebrand, K. G. 1978. Labour and capital in the Scandinavian countries in the nineteenth and twentieth centuries. In P. Mathias and M. M. Postan, eds., The Cambridge Economic History of Europe: The Industrial Economies—Capital, Labour and Enterprise: Britain, France, Germany and Scandinavia. Cambridge: Cambridge University Press.

Hitch, C. J., and R. N. McKean. 1960. The Economics of Defense in the Nuclear Age. Cambridge, Mass.: Rand Corporation and Harvard University Press.

Hodenheimer, A. J. 1978. Major Reforms of the Swedish Education System 1950-1975. Washington, D.C.: World Bank Staff Working Paper No. 290.

Hogwood, B. W., and B. G. Peters. 1985. The Pathology of Public Policy. Oxford: Clarendon Press.

Imai, K., and H. Itami. 1981. The Firm and Market in Japan—Mutual Penetration of the Market Principle and Organization Principle. Tokyo: Institute of Business Research Discussion Paper No. 104, Hitotsubashi University, June, mimeo.

Imai, K., I. Nonaka, and H. Takeuchi. 1984. Managing the New Product Development Process: How Japanese Companies Learn and Unlearn. Tokyo: Institute of Business Research Discussion Paper No. 118, Hitotsubashi University, March 29, mimeo.

Johnson, C. 1982. MITI and the Japanese Miracle—The Growth of Industrial Policy 1925-1975. Stanford: Stanford University Press.

Jones, I., and H. Hollenstein. 1983. Trainee Wages and Training Deficiencies: An Economic Analysis of a "British Problem." London: National Institute of Economic and Social Research Industry Series No. 12, June, mimeo.

Kamien, M. I., and N. L. Schwartz. 1982. Market Structure and Innovation. Cambridge: Cambridge University Press.

Katz, B. C., and A. Phillips. 1982. The computer industry. In R. Nelson, ed., Government and Technical Progress. New York: Pergamon Press.

Katzenstein, P. J. 1985a. Corporatism and Change: Austria, Switzerland, and the Politics of Industry. Ithaca, N.Y.: Cornell University Press.

Katzenstein, P. J. 1985b. Small States in World Markets: Industrial Policy in Europe. Ithaca, N.Y.: Cornell University Press.

Kono, T. 1984. Strategy and Structure of Japanese Enterprises. London: Macmillan.

Kreps, D. M., and A. M. Spence. 1985. Modelling the role of history in industrial organization and competition. In Feiwel, ed., Issues in Contemporary Microeconomics and Welfare Analysis. Cambridge: Cambridge University Press.

Laboratorio di Politica Industriale. 1982. Materiali de Discussione. Bologna: Laboratorio di Politica Industriale, November, mimeo.

Lafay, G. 1985. Spécialization française: Des handicaps structurels. Revue d'Economie Politique 95(5).

Launer, H., and W. Ochel. 1985. Industrielle strukturanpassung: Das Japanische modell. Ifo-Schelldienst (September 26).

Lavington, S. 1980. Early British Computers: The Story of Vintage Computers and the People Who Built Them. Manchester: The Digital Press.

Leamer, E. E. 1984. Sources of International Comparative Advantage: Theory and Evidence. Cambridge, Mass.: MIT Press.

Le Monde. 6 February 1986. M. Jean-Jacques Duby quitte le CNRS.

Le Monde. 27 September 1979. Le rapport hannoun souligue la forte concentration et la faible efficacité des aides publiques à l'industrie.

Liebenau, J. 1985. Innovation in pharmaceuticals: Industrial R&D in the early twentieth century. Research Policy 14 (August).

Lipsey, R. E., and I. B. Kravis. 1985. The Competitive Position of U.S. Manufacturing Firms. Cambridge, Mass.: National Bureau of Economic Research Working Paper No. 1557, February.

Maclaurin, W. R. 1949. Invention and Innovation in the Radio Industry. New York: Macmillan.

Malerba, F. 1985. Demand structure and technological change: The case of the European semiconductor industry. Research Policy 14 (October).

Maurice, M., F. Sellier, and J.-J. Sylvestre. 1982. Politique d'Education et Organisation Industrielle en France et en Allemagne. Paris: Presses Universitaires de France.

Meyer-Krahmer, F., G. Gielow, and U. Kuntze. 1983. Impacts of government incentives towards industrial innovation. Research Policy 12 (June).

Milward, A. S., and S. B. Saul. 1977. The Development of the Economies of Continental Europe 1850-1914. Cambridge, Mass.: Harvard University Press.

Mitchell, J. P. 1977. New Directions for Apprenticeship Policy. Worklife (January). Washington, D.C.: U.S. Department of Labor.

Mowery, D. C. 1983a. Economic theory and government technology policy. Policy Sciences 12 (August).

Mowery, D.C. 1983b. Innovation, market structure, and government policy in the American semiconductor electronics industry: A survey. Research Policy 12 (August).

Mueller, D. 1985. Persistent Performance Among Large Corporations. Brussels: Centre for European Policy Studies, November, mimeo.

Murray, C. 1984. Losing Ground: American Social Policy 1950-1980. New York: Basic Books.

Nakamura, T. 1981. The Postwar Japanese Economy—Its Development and Structure. Tokyo: University of Tokyo Press.

Nakayama, S. (J. Dusenbury, trans.). 1984. Academic and Scientific Traditions in China, Japan and the West. Tokyo: University of Tokyo Press.

National Academy of Engineering. 1982. The Competitive Status of the U.S. Auto Industry. Committee on Technology and International Economic and Trade Issues, Automobile Panel. Washington, D.C.: National Academy Press.

National Manpower Council. 1954. A Policy for Skilled Manpower. New York: Columbia University Press.

National Science Board. 1986. Science Indicators—The 1985 Report. Washington, D.C.: National Science Foundation.

National Science Foundation. 1985. Science and Engineering Personnel: A National Overview. Washington, D.C.: National Science Foundation.

Nelson, R. R., ed. 1982. Government and Technical Progress: Cross-Industry Analysis. New York: Pergamon.

Nelson, R. R. 1984. High-Technology Policies: A Five Nation Comparison. Washington, D.C.: American Enterprise Institute.

Nelson R. R., and S. G. Winter. 1982. An Evolutionary Theory of Economic Growth. Cambridge, Mass.: Harvard University Press.

Noble, D. F. 1977. America by Design. Oxford: Oxford University Press.

Nomura Research Institute. 1983. Characteristics of Japan's Import and Export Structures. Tokyo: Nomura Research Institute.

Nonaka, I., T. Kagono, and S. Sakamoto. 1983. Evolutionary Strategy and Structure—A New Perspective on Japanese Management. Tokyo: Institute of Business Research Discussion Paper No. 111, Hitotsubashi University, March, mimeo.

Nove, A. 1983. The Economics of Feasible Socialism. London: George Allen & Unwin.

Office Fédéral de l'Industrie, des Arts et Métiers et du Travail. 1980. Politique Concernant le Marché du Travail en Suisse: Caractéristiques et Problèmes, Vol. 1. Berne.

Ohlsson, L. A. 1980. Engineering Trade Specialization of Sweden and Other Industrial Countries. Amsterdam: North Holland Publishing Co.

Nelson, R. R., ed. 1982. Government and Technical Progress: Cross-Industry Analysis. New York: Pergamon.

Nelson, R. R. 1984. High-Technology Policies: A Five Nation Comparison. Washington, D.C.: American Enterprise Institute.

Nelson R. R., and S. G. Winter. 1982. An Evolutionary Theory of Economic Growth. Cambridge, Mass.: Harvard University Press.

Noble, D. F. 1977. America by Design. Oxford: Oxford University Press.

Nomura Research Institute. 1983. Characteristics of Japan's Import and Export Structures. Tokyo: Nomura Research Institute.

Nonaka, I., T. Kagono, and S. Sakamoto. 1983. Evolutionary Strategy and Structure—A New Perspective on Japanese Management. Tokyo: Institute of Business Research Discussion Paper No. 111, Hitotsubashi University, March, mimeo.

Nove, A. 1983. The Economics of Feasible Socialism. London: George Allen & Unwin.

Office Fédéral de l'Industrie, des Arts et Métiers et du Travail. 1980. Politique Concernant le Marché du Travail en Suisse: Caractéristiques et Problèmes, Vol. 1. Berne.

Ohlsson, L. A. 1980. Engineering Trade Specialization of Sweden and Other Industrial Countries. Amsterdam: North Holland Publishing Co.

Organization for Economic Cooperation and Development. 1979. Policies for Apprenticeship. Paris: Organization for Economic Cooperation and Development.

Organization for Economic Cooperation and Development. 1984a. Industry and University: New Forms of Co-operation and Communication. Paris: Organization for Economic Cooperation and Development.

Organization for Economic Cooperation and Development. 1984b. Mergers and Takeovers. Paris: Organization for Economic Cooperation and Development.

Organization for Economic Cooperation and Development. 1986. Changes in Working Patterns and the Impact on Education and Training: Human Resources Policies and Strategies in Germany. Paris: Organization for Economic Cooperation and Development.

Orléan, A. 1986. L'insertion dont les Échanges Internationaux. Economie et Statistique (January).

Peck, M., and R. Wilson. 1982. Innovation, imitation, and comparative advantage. In H. Giersch, ed., Emerging Technologies: Consequences for Economic Growth, Structural Change and Employment. Tübingen: J. C. B. Mohr.

Peterson, T., and F. J. Comes. 1985. An electronics dream that's shorting out. Business Week (March 4).

Pham-Khac, K., and J. L. Pigelet. 1979. La Formation et l'Emploi des Docteurs ès Sciences. Dossier du Centre d'Etudes et de Recherches sur les Qualifications. Paris.

Phillips, A. 1971. Technology and Market Structure: A Study of the Aircraft Industry. New York: Heath Lexington Books.

Picard, J.-F., A. Beltran, and M. Bungener. 1985. Histoire(s) de l'EDF: Comment se Sont Prises les Décisions de 1946 à Nos Jours. Paris: Bordas.

Ponssard, J. P., and G. de. Pouvoirville. 1982. Marché Publique at Innovation. Paris: Economica.

Prais, S. J. 1981. Productivity and Industrial Structure. Cambridge: Cambridge University Press.

Prais, S. J., and K. Wagner. 1983a. Schooling Standards in Britain and Germany: Some Summary Comparisons Bearing on Economic Efficiency. London: National Institute Discussion Paper No. 60.

Prais, S. J., and K. Wagner. 1983b. Some practical aspects of human capital investment:

Training standards in five occupations in Britain and Germany. National Institute Economic Review (August).

Pratten, C. F. 1976. A Comparison of the Performance of Swedish and U.K. Companies. Cambridge: Cambridge University Press.

Research & Planning Institute, Inc. 1980. Case Studies Examining the Role of Government R&D Contract Funding in the Early History of High Technology Companies. Cambridge, Mass.: Research & Planning Institute, Inc.

Riche, R. W., D. E. Hecker, and J. U. Bergan. 1983. High technology today and tomorrow: A small slice of the employment pie. Monthly Labor Review (November).

Robson, M., J. Townsend, and K. Pavitt. 1985. Sectoral Patterns of Production and Use of Innovations in the UK: 1945-1983. Centre for Science, Technology and Energy Policy (Economic and Social Research Council), May 30, mimeo.

Rosenberg, N., and L. E. Birdzell, Jr. 1986. How the West Grew Rich: The Economic Transformation of the Industrial World. New York: Basic Books.

Rosenberg, N. 1976. Perspectives on Technology. Cambridge: Cambridge University Press.

Rotemberg, J. J., and G. Saloner. 1984. A Supergame-Theoretic Model of Business Cycles and Price Wars During Booms. Cambridge, Mass.: MIT Working Paper No. 349, July, mimeo.

Rothwell, R., and W. Zegveld. 1981. Industrial Innovation and Public Policy. Westport, Conn.: Greenwood Press.

Ryan, P. 1984. Job training, employment practices and the large enterprise: The case of costly transferable skills. In P. Osterman, ed., Internal Labor Markets. Cambridge, Mass.: MIT Press.

Saxonhouse, G. R. 1984. What is all this about "industrial targeting" in Japan? World Economy (September).

Sandberg, L. G. 1979. The case of the impoverished sophisticate: Human capital and Swedish economic growth before World War I. The Journal of Economic History 39 (March).

Scherer, F. M. 1982. Inter-industry technology flows in the United States. Research Policy 11 (August).

Scherer, F. M., and D. Ravenscraft. 1984. Growth diversification: Entrepreneurial behavior in large-scale United States enterprises. Zeitschrift für Nationalökonomie (Suppl. 4).

Schmalensee, R. 1982. Product differentiation advantages of pioneering brands. American Economic Review 72 (June).

Schröder-Gudehus, B. 1972. The argument for the self-government and public support of science in Weimar Germany. Minerva.

Sciences and Technology Agency (Japan). 1985. White Paper on Science and Technology 1986. Tokyo: Foreign Press Center of Japan.

Scribberas, E. et. al. 1978. Competition, Technical Change and Manpower in Electronic Capital Equipment: A Study of the U.K. Mini-computer Industry. Brighton: Science Policy Research Unit.

Shinshara, M. 1970. Structural Changes in Japan's Economic Development. Tokyo: Konokuniya.

Shonfield, A. 1965. Modern Capitalism. Oxford: Oxford University Press.

Stevens, B. 1986. Labour markets, education and industrial structure. In H. Ergas, ed., A European Future in High Technology? Brussels: Centre for European Policy Studies.

Stoffaes, C. 1984. Politique Industrielle. Paris: Les Cours de Droit.

Sturmes, S. G. 1958. The Economic Development of Radio. London: Duckworth.

Suzuki, K. 1985. An Empirical Analysis of the Interdependence of R&D Investment and

Market Structure in Japan. Tokyo: Research Institute of Capital Formation, The Japan Development Bank Staff Paper No. 4, October, mimeo.

Tachibanki, T. 1984. Labour mobility and job tenure. In M. Aoki, ed., The Economic Analysis of the Japanese Firm. Amsterdam: North-Holland.

Teubal, M., and E. Steinmueller 1982. Government policy, innovation and economic growth. Research Policy 11 (October).

Uena, H. 1977. Conception and evaluation of Japanese industrial policy. Japanese Economic Studies (Winter 1976-1977).

Vellas, F. 1981. Echanges Internationaux et Qualification du Travail. Paris: Economica.

Vernon, R. 1966. International investment and international trade in the product cycle. Quarterly Journal of Economics 80 (May).

Vernon, R. 1974. Big Business and the State: Changing Relations in Western Europe. London: Macmillan.

Vernon, R. 1979. The product cycle hypothesis in a new international environment. Oxford Bulletin of Economics and Statistics 41 (November).

Webbink. D. W. 1977. The Semiconductor Industry. Washington, D.C.: Government Printing Office.

Weinberg, A. M. 1967. Reflections on Big Science. Oxford: Pergamon Press.

Wilson, R. A. 1980. The rate of return to becoming a qualified scientist or engineer in Great Britain, 1966-1976. Scottish Journal of Political Economy (February):41-62.

Worswick, G. D. N., ed. 1985. Education and Economic Performance. Aldershot, England: Gower Publishing Company.

Young, S., and A. V. Lowe. 1974. Intervention in the Mixed Economy. London: Croom Helm.

National and Corporate Technology Strategies in an Interdependent World Economy

LEWIS M. BRANSCOMB

The broad public perception that important conflicts occur between the sovereignty of nations and the business interests of transnational corporations can be illustrated by a lecture given a quarter century ago by Tony Wedgewood Benn—first U.K. Minister of Technology. Speaking at the British Embassy in Washington, D.C., he introduced the idea that Her Majesty's government should appoint ambassadors to the largest multinational companies—on the grounds that the companies' gross revenues exceed the gross national product of many small countries and that their policies had more impact on U.K. interests than did those of small states.

In introducing the metaphor of foreign relations as a means for dealing with significant external forces with which one must cope, Mr. Benn was reflecting the presumption of most governments that multinational operations in their countries pose a threat that has to be defended against, rather than an opportunity that could be exploited to the countries' advantage. Either way, of course, foreign ministries are clearly not the most competent instruments of government to deal with technological issues.

In his modest proposal Mr. Benn also caricatured the notion of sovereignty, which, of course, refers to the authority vested in a nation-state to exercise control over its own territories. His humor anticipated a certain amount of political agitation that exists today over these conflicts. In fact, however, when analyzed more closely, the fundamental interests of nations and corporations are surprisingly similar. And learning together to balance national activities and international dependencies is the key to relieving

246

current stresses and to achieving both parties' long-term economic and social interests.

THE DIFFERENT RESPONSIBILITIES OF
GOVERNMENTS AND CORPORATIONS

During the nineteenth century, great mercantile firms such as the British East India Company—with either tacit or explicit support from their governments—did indeed challenge sovereign authority in many parts of the globe. During the declining phases of colonialism in that century, news stories persisted of banana companies, such as United Fruit, manipulating corruptible Central American governments. This experience of many developing countries emerging from their colonial past still colors national attitudes today. It makes such governments hypersensitive to maintaining their sovereignty unsullied by either military or economic challenge.

Very few matters of current corporate interest in industrialized democracies result in direct confrontational challenges to sovereignty by companies. When the governments of India, Nigeria, or Indonesia decide to adopt policies that IBM believes reduce its prospects of success below acceptable levels, the company does not hire mercenaries; it leaves the country or changes the way it does business. Though direct corporate challenges to national sovereignty are disappearing, there is, it seems, a concomitant growth in the ways in which an increasingly global economy impinges on sovereignty.

At the heart of the friction between interdependent nations and corporations is a simple truth: Although neither companies nor nations are absolute masters of their fate, companies have a great adaptive advantage in their ability to redefine their commitments in response to opportunity or to changes in their political environment. Governments' responsibilities, on the other hand, are less flexible. Governments are responsible to and for individuals living within their national borders, are directed by a constitution or set of political norms that specify relatively inflexible commitments, and are both burdened and inspired by the blood, suffering, and heroic myths invested in the defense of their physical boundaries.

The issue of interest, therefore, is not companies challenging sovereignty, but a more complex question of synergy or conflict between the interests of two types of global organizations. Both nations and companies have interests that cross physical borders, but nations are committed to a particular piece of geographic "turf" and companies are not. The complexity of interaction between these two types of organizations is rising fast for both political and technological reasons.

Technology diffuses faster and further than ever before, and most of the effective transfer of technology is planned and accomplished by corporations. However, no nation can leave the acquisition of technological capability to chance. Governments not only engage in the direct promotion of technology but increasingly seek to influence the strategies of enterprises. They have a keen interest in the success of technology-promotion activities, even when prospects for success are low, but businesses are unlikely to invest if success is unlikely, unless the potential rewards are commensurate. Companies choose their own strategies to match expectations of market opportunity to investment risk, while accommodating the policies of governments in each country where they choose to do business.

The current situation in the People's Republic of China (PRC) is a good illustration of these issues, because a unique high-risk/high-reward situation exists there for both the government and foreign business. Businesses starting joint ventures in China may be pessimistic about their likely commercial success, but the potential markets are so huge that, in the long run, even a small chance of success warrants a considerable effort and investment. The PRC's need to master certain key technologies also creates a high-risk/high-reward situation, in that the importance of the technology may warrant considerable political risk in trying to acquire it through joint ventures with foreign firms, even when this might give a particular firm undesired market power in the country or lead to unwanted foreign political or cultural influences. As illustrated by the next section, trade-offs are less clear in other situations.

GOVERNMENTAL POLICIES AFFECTING TECHNOLOGY

Just as tolerance for risk and expectations for success are sometimes perceived differently by businesses and political bodies, motivations also may differ greatly. Trade policy and national security concerns dominate U.S. government interest in corporate technology strategy. The expansion of federal R&D investment is now largely in defense. Encouraging friendly governments to participate in the Strategic Defense Initiative (SDI) coexists uneasily in the U.S. technology policy agenda with export controls on technology and access to foreign markets for U.S. "high-tech" companies.

At the same time, both in the United States and abroad, there is concern that the country not become dependent for strategic goods on a supplier in a foreign nation that might not be reliable under certain future political conditions. All countries naturally feel the need to preserve political control over their defense industrial base, because future alignments of their suppliers of key components for military purposes are unpredictable.

In the case of SDI, the United States invites European participation to encourage political support for the program in return for a certain amount of technology sharing. But, on the other hand, the U.S. government does not want to increase the chances of leakage of the militarily relevant pieces of the technology to the Eastern Bloc. Reducing budgetary costs through cost sharing is also an important motivation, but the effort is to achieve the maximum amount of cost sharing and political solidarity behind SDI without sharing more technology than is absolutely necessary.

Many Europeans, for their part, fear that failure to participate in SDI might disadvantage their economies if the SDI technologies turn out to stimulate U.S. commercial competitiveness. They harbor the suspicion that U.S. reticence to share fully is based at least as much on reluctance to accept European industrial parity as it is on fear that technology will leak to the Soviets.

The same considerations are apparently operating in negotiations over European and Japanese cost sharing and cooperation in the Space Station program. Here again the U.S. objective appears to be to achieve a foreign financial contribution without sharing any more critical technology than is necessary. Expectations at present appear unrealistic in this regard.

Japanese policy, on the other hand, is focused on industrial competitiveness in export markets as the primary strategy for dealing with Japan's dependence on imported food, energy, and raw materials. Industry has a strong voice in the government's technology strategy, and government concerts industry action in the export arena, leaving domestic competition largely free of government direction.

In Europe there is superposed on each country's national policy a regional attempt to gain the benefits of European economic integration without paying the political and cultural penalty of more extensively shared sovereignty. The success of U.S. and Japanese firms in the strategic high-tech industries and concern about European competitiveness are motivating a variety of national and European government programs to encourage collaboration among firms and universities in research of strategic commercial importance. The extent to which national companies and the national subsidiaries of transnational enterprises seek—and are allowed—to participate in these programs provides an interesting test of the compatibility of corporate and governmental technology policies.

The People's Republic of China offers a particularly interesting example because public policy for technology promotion is in such a rapid state of change. The liberalized economic policy of the current government seeks to leverage foreign-owned technology through encouraging foreign investment under negotiated terms and thereby attain technological self-sufficiency as quickly as possible. Corporate and national strategies for technology mastery by indigenous Chinese enterprises are explicit and often central elements in that negotiation.

A final example is Brazil, which follows a mixed strategy, centrally directed, to encourage foreign investment in selected areas and reserve the large domestic market in others. This market-reservation policy is pursued with considerable tenacity, even to the point of refusing permission to foreign firms willing to export from Brazil the majority of their manufactured output. The conflict between the interests of firms and the policies of government is highly visible in this situation, since there is such deep disagreement on the effectiveness—and risk—of this strategy for technical development.

It seems clear that Brazil is trying to use the protected, elevated price structure derived from reservation of its domestic market to finance learning-by-doing in technology, regarded as key to the country's economic future. The government believes this will produce the most rapid feasible achievement of self-sufficiency in technology. In effect, domestic customers are being asked to subsidize the self-education of local technologists, without the stimulation of external competition.

The underlying assumption is apparently that this costly learning-by-doing will be more effective in the long run than assimilation of foreign technology through joint ventures and foreign direct investment. This strategy contrasts with that of the PRC, which apparently believes that more rapid technology transfer through carefully controlled foreign investment more than offsets the risk that technology assimilation by this mechanism will be superficial and ineffective.

These examples illustrate the diverse motivations and approaches to technology policy around the world. In each country, the national enterprises seek the support of their government while seeking freedom of action within its policies. But, in most cases, conflicts of interest between transnational companies and the central government do not turn on challenges to the sovereign authority of government. Rather, they hinge on the struggle for domestic political consensus on economic and industrial policy and debates about sharing those policies with other nations. Therefore, the following discussion focuses on the subsidiaries of transnational companies and their relations with host governments and national enterprises.

Although the interests of governments and transnational corporations differ, it should be noted that they share many—indeed most—objectives in common. Companies are often eager to invest in building high-tech capability in countries where local markets are accessible and need local support capabilities. Political leaders in such countries are often equally eager to accelerate technological development in the search for new jobs and productivity and to ease industrial readjustment. In many countries the subsidiaries of transnationals enjoy a special relationship with government close to that enjoyed by domestically based transnationals, even to the extent of substantial participation in joint ventures, government-funded enterprises, and sales to government agencies.

The IBM Corporation's experience in Europe and in Japan has been very positive, despite occasionally publicized controversies. IBM Japan is a wholly owned subsidiary of IBM, dating back to the 1930s, which participates effectively in Japanese and Asian markets, making a substantial positive contribution to Japanese exports to Asia and South America. IBM Italy is officially designated a national Italian company because of its contributions to the country's employment, economy, and development of science, technology, and professional skills.

SOURCES OF CONFLICT

Given this synergy of interest, why do the inevitable divergences in technology policies of governments and transnational business attract so much attention and occasionally cause so much friction? Conflict between the interests of transnational companies and those of national governments arises from several chief sources. First, transnational companies are sometimes better positioned to take advantage of policies for enhancing regional competitiveness than are national companies. When the European Economic Community (EEC) sought to integrate the European economies, firms such as IBM had little difficulty organizing their businesses to take advantage of the special strengths of each country, avoid needless duplication, and serve the entire market with a product line of minimum national diversity. IBM's success demonstrates the effectiveness of the EEC strategy, provided there are private firms willing and able to take advantage of it.

Transnationals based in Europe have been much slower to find those advantages, which are equally available to all. Indeed, in 1970 three governments urged three of their domestically based multinationals—Siemens, CII-Honeywell-Bull, and Philips—to organize a joint marketing venture called Unidata to aggregate a Europewide market for their computers. It soon fell apart when the Dutch withdrew and the French consortium encountered hard times. Commercial marriages made through political efforts are vulnerable to failure if the natural forces of business motivation are weak or absent.

A second source of conflict is the inevitable political perception that any strategy that increases economic dependence on foreign enterprises and their governments necessarily reflects a diminution of control over the nation's destiny. There are plenty of signs of increasing U.S. sensitivity to dependence on foreign suppliers. Indeed, the defense market in the United States has tended to remain a highly protected market through buy-American laws and other policy interventions to ensure that U.S. military capability is based primarily on a domestic industrial base, even at high incremental cost. This is especially evident, for example, in shipbuilding.

Overall, however, Americans are still much less sensitive to the political burden of an ever-increasing dependence on trade for national survival. Europeans and Japanese have lived with this reality for a very long time. We will learn to live with it too as our dependence grows. It is to be hoped that when our dependence equals theirs, our government's chauvinism in dealing with foreign enterprises is no worse than theirs.

Third, all governments must deal with the realities of employment, inflation, and growth, and also with public perceptions of the nation's ability to improve its prospects. In most democracies those public perceptions, as amplified by the media, are the politically relevant measure of the effectiveness of government policies. Indeed, the more severe the challenges of unemployment and inflation, the more conspicuous is the public concern about foreign dependence and control.

Everyday policies are often pragmatic, designed to encourage both foreign and domestically owned enterprises to invest and expand the economy. But structural economic change is hard to measure objectively; the perception of change may be politically much more salient than the reality. Thus, most governments give great political weight to the significance of domestic ownership of national enterprises. When acting in the public view, government officials may favor a domestically owned company that imports goods of foreign manufacture over a transnational firm's local subsidiary, even though it employs thousands of people; invests heavily in training, research, and development; and contributes more to national economic performance and the balance of trade than the domestic firm. Local ownership is, therefore, a symbol of national self-sufficiency, but it is often a poor criterion for economic development policy.

Fourth, conflict can arise when governments, including that of the United States, assert the right to extend national sovereignty to domestic firms' subsidiaries on foreign soil and thereby exacerbate nationalistic political reactions. IBM's ability to market its largest computers to French government agencies is still adversely affected by French memories of the U.S. government's refusal to license large IBM computers to the French atomic energy authority as a point of pressure on Charles de Gaulle over French nuclear policy. These old memories have been revived by current processes for licensing large vector processors to Western Europe. The Caterpillar Tractor Company has only recently recovered from the disruption of its European business by the U.S. government's intervention in contract compliance on the Europe-USSR pipeline. In the meantime, Japanese competitors have greatly increased their market share in this business sector.

Finally, political leaders of countries experiencing economic difficulties, or nurturing nascent industries, may also discriminate against foreign investment even when it builds local infrastructure, reduces unemployment, and helps solve serious balance-of-trade problems. National subsidiaries, in response, attempt to demonstrate in every way they can that they are, in fact, national companies and make positive contributions to national well-being.

The more enlightened companies recruit and train nationals to manage and staff their local subsidiaries. They work hard to integrate the subsidiary into the social, economic, and technical life of the community. Yet, they too must balance this emphasis on national identity with the extranational interests of the enterprise of which they are a part. They are often torn between the imperatives of a global strategy and the perceived priorities of local national economic development.

NEW TECHNOLOGICAL STRATEGIES

Companies as well as governments share an increasingly complex task. The world is evolving from a loosely coupled set of independent nation-states to a highly complex world economy whose institutions of extra- and international sovereignty are still undeveloped, and in which national boundaries are often a poor fit to more natural economic regions.

Reflecting their awareness of these political realities, managements of both transnational companies and governments pursue technological strategies that are designed in part to reinforce perceptions that are important to them. Companies, for example, try to locate their manufacturing facilities to minimize negative balance-of-trade impacts and to demonstrate a significant contribution to national technological capability. Nations, alone or in regional concert, embark on highly visible technology projects in the hope of stimulating industrial collaboration, building consensus around long-term, technology-intensive industrial strategies.

The long series of 10-year projects sponsored by the Ministry of International Trade and Industry (MITI) in Japan, perhaps the best-known recent example of such a long-term industrial strategy, stimulated much of the enthusiasm for Europe's response in similar vein. Conditions in culturally homogeneous Japan, however, are much more propitious for cooperative technology ventures on a national scale than the conditions for regional cooperation in Europe. In addition, many people exaggerate the importance of technical projects supported by MITI, which are more a symptom of a nationally concerted strategy for capturing a larger share of world markets in selected industries than the source of that trade success. Nevertheless, there is no doubt that, even though some of the MITI programs (such as Pattern Information Processing Systems, or PIPS) are not regarded as successful by many, the Japanese government does direct a substantial amount of funding into carefully selected, commercial technologies of perceived future importance. In most programs, although surprisingly not in its current artificial intelligence program at the Institute for New Generation Computer Technology (ICOT), the government gives strong emphasis to field trials and applications research to test market acceptance.

In Europe, the EEC sponsors the European Strategic Program for Research and Development in Information Technology (ESPRIT), which includes a large number of relatively small research-oriented projects at the precompetitive level. The most commercially important of these projects is probably the collaboration between Siemens and Philips in submicron integrated circuit technology. IBM does participate in ESPRIT. After careful and protracted negotiations, the IBM Europe/Middle East/Africa Corporation was allowed to submit proposals, two of which have been accepted and are under way. One, for example, deals with the application of artificial intelligence techniques to computer-integrated manufacturing. The breadth of participation in ESPRIT is probably due primarily to the availability of EEC funds, representing "new money" that is less subject to parochial national interests than the funds of national governments would have been and also the precompetitive nature of the projects.

A second EEC program, called RACE (Research in Advanced Communications in Europe), funds cooperative projects in digital telecommunications, anticipating the broad-band, integrated voice and data networks of the future. In this case, transnational companies share with the EEC the hope that the RACE project will put pressure on the national post, telephone, and telecommunications authorities to adopt more ambitious and compatible technical goals. The balkanization of the telecommunications infrastructure of Europe is a significant handicap, to both technological and economic modernization. Whether a set of cooperative technology projects such as RACE can achieve this result is, of course, an open question, but in this case national, EEC, and company interests— both European and foreign based—strongly coincide.

Another example is the Community in Education and Training for Technology (COMETT) program, in which advanced satellite-based industrial education experience will be shared. IBM is willing to make an important contribution to COMETT, and it is highly likely that this will be welcomed.

In the United States, government-initiated projects to promote commercial strategies may have their origin in response to military R&D initiatives. As noted earlier, in 1985 the United States put considerable political pressure on European governments to participate in the vastly expensive and ambitious research and technology development for the SDI. The European reaction, especially in France, was to accept the proposition that spin-off technology from SDI might give them commercial advantage in the future (or at least prevent them from experiencing a further commercial disadvantage relative to the United States), yet to fear that U.S. export controls would be used later to limit European freedom to use the technology commercially if they did participate.

President Mitterand's response was to launch the European technology cooperation program called EUREKA, whose projects are chosen for their commercial, not military, value and unlike ESPRIT are not limited to precompetitive research.

European governments have embraced EUREKA in principle and officially regard it as "complementary" to the EEC's "Technology Community" proposal. Like EUREKA and ESPRIT, the Technology Community proposal focuses on the set of technologies that all countries generally regard as strategic: informatics, biotechnology, advanced materials, automation, telecommunications, lasers, and educational technologies.

A mechanism is in place for selecting EUREKA projects, and 26 have been selected. But there is no reserved funding at the EEC or intergovernmental level for EUREKA projects. They are referred back to national governments for possible support. Thus, it is not surprising that private enthusiasm for EUREKA does not match the political support it enjoys in public, indicating that private assessment of the commercial benefits is considerably less optimistic than the public assessment.

Nevertheless, there is a broad movement, in Europe and Japan especially, to put high on the political as well as the business agenda the quest for what Hubert Curien, French Minister of Research and Technology, called the "technical renaissance of Europe." The leaders of companies and governments alike must surely welcome this new priority, even as they debate what forms of public action will produce the desired results.

Since such highly visible national or multilateral projects require the participation of commercial firms across national lines, difficulties do arise at this juncture. Firms, like governments, appreciate the symbolic importance of technological success, for they too have a vital interest in public perceptions. But firms measure the results of their investments by more conservative criteria than do governments: high expectation of a contribution to competitive success.

Thus, the prevailing pattern is that firms publicly support the goals of the government projects and will generally seek to participate, especially if their competitors do. The projects are generally not only "preproprietary" but are smaller than mainstream corporate projects. The occasionally much larger projects are sometimes joint ventures already under negotiation among private firms.

Both transnational companies and governments must manage the balance between national activities and international interdependences. A company's subsidiaries in different countries are collaborating, not competing, with each other. The company has an incentive to build its technical strategy on the totality of its worldwide capability. A government more typically regards the national economy in competition with all others, and its leaders seek to optimize the economic advantages of their electorates, independent of the others.

A problem here is that "economic advantage" tends to be defined as benefits to producers rather than to consumers, who are much less well organized to protect and register their interests. It is not at all clear that, even in the narrowest terms, the quality of life of electorates overall is improved by nationalistic strategies.

Of course, since national subsidiaries work hard at earning acceptance as national entities, they too share national economic aspirations, and their political views may well be in full accord with their governments'. In most regions of the world, both political and economic realities are forcing nations to make common cause at least regionally, and the European Common Market is the most conspicuous example.

Thus, the tensions between self-sufficiency and interdependence, between nationalism and global development, between technology for profit and technology for investment in public perception, between strategies seeking market access and political acceptance and strategies optimizing near-term total business performance, will continue. Eventually, the international diffusion of culture and ethnicity, and the imperatives of global interdependence, will begin to relieve these stresses and soften the more strident nationalistic trends. But it will take a long time. Meanwhile, we must work toward trade policies, international technological exchange, and international cooperation that is based more on economic realities that nations and companies face than on time-honored symbols of sovereignty that may not accord with anyone's long-term national interest.

ADVISORY COMMITTEE FOR SYMPOSIUM ON WORLD TECHNOLOGIES AND NATIONAL SOVEREIGNTY

Chairman

HARVEY BROOKS, Benjamin Peirce Professor of Technology and Public Policy, Emeritus, Harvard University

Members

ANN F. FRIEDLAENDER, Dean, School of Humanities and Social Science, Massachusetts Institute of Technology

FREDERICK W. GARRY, Vice President, Corporate Engineering and Manufacturing, General Electric Company

HENRY KRESSEL, Managing Director, Warburg, Pincus and Company

JACK D. KUEHLER, Senior Vice President, IBM Corporation

JAMES BRIAN QUINN, William and Josephine Buchanan Professor of Management, Amos Tuck School of Business, Dartmouth College

RAYMOND VERNON, Clarence Dillon Professor of International Affairs Emeritus, Kennedy School of Government, Harvard University

Contributors

LEWIS M. BRANSCOMB is Professor from Public Service at Harvard University and director of the Science, Technology, and Public Policy Program at the John F. Kennedy School of Government. Dr. Branscomb came to Harvard University in October 1986 from the IBM Corporation, where he was vice president and chief scientist and a member of the Corporate Management Board.

HARVEY BROOKS is Benjamin Peirce Professor of Technology and Public Policy Emeritus at Harvard University and a member of the National Academy of Engineering and the National Academy of Sciences.

YVES DOZ is associate professor of business policy at the European Institute for Business Policy. His current research interests cover the management of innovation in large, complex firms, with emphasis on information technologies and pharmaceutical industries.

HENRY ERGAS is counsellor in the Advisory Unit to the Secretary-General at the Organization for Economic Cooperation and Development (OECD), in Paris. He joined the OECD in 1978 and has mainly been responsible for work in the areas of telecommunications, industrial, and trade policy.

BRUCE R. GUILE is associate director of the Program Office of the National Academy of Engineering.

ALVIN P. LEHNERD is executive vice president, North American Appliance

Group, Allegheny International. Before joining Allegheny International in 1982, Mr. Lehnerd spent 13 years with the Black & Decker Manufacturing Company, where his last position was as vice president for advanced technology, new business, and new ventures.

JAMES BRIAN QUINN is William and Josephine Buchanan Professor of Management at the Amos Tuck School of Business Administration, Dartmouth College. Professor Quinn is a consultant to leading U.S. and foreign companies, the United States and foreign governments, and a number of small enterprises. He has published extensively on strategic planning, entrepreneurship, and technology policy and has served on numerous committees for the National Research Council and the National Academy of Engineering.

DAVID J. TEECE is professor of business administration at the School of Business and director of the Center for Research in Management at the University of California, Berkeley. He earned his Ph.D. in economics at the University of Pennsylvania and joined the faculty of the Stanford Business School in 1975. Professor Teece came to the University of California in 1982.

JAMES M. UTTERBACK is the director of the Industrial Liaison Program of the Massachusetts Institute of Technology and an associate professor in MIT's School of Engineering. Since receiving his Ph.D. in the MIT Sloan School's technology management program, Professor Utterback has held faculty positions at Indiana University and the Harvard Business School.

RAYMOND VERNON is Clarence Dillon Professor of International Affairs Emeritus at Harvard University. He was formerly the director of Harvard's Center for International Affairs, as well as director of the Multinational Enterprise Study of the Harvard Business School.

Index

A

Aircraft and airlines industries,
  innovations in, 34
  imitators profiting from, 67
  standardization of, 74
Antitrust policies, affecting international
    competitiveness of United States,
    174-175
Apprenticeship system, 208
Appropriability of innovations, 67-68,
    89-90
  access to complementary assets
    affecting, 70-72, 77, 80, 89-90
  industry differences in, 89-90
  and profitability, 67-68, 72-76
  in international trade, 90-93
Assets required for commercialization of
    innovations, 21, 23, 26, 39-40,
    70-72, 76-88
  contractual strategies concerning, 73-
    74, 77-79
    combined with integration
      strategies, 85

compared to integration strategies,
    80-85
  integration strategy for access to 76-
    77, 79-85; *see also* Integration
    strategy
  loose, 68, 74, 80, 81, 82, 90-91
  parterning strategy concerning, 3, 78-
    79, 90, 102, 103, 107, 115
  and profitability, 67-68, 72-76
    in international trade, 90-93
  specialized, 71, 72, 75-76, 84
  strategies for access to, 76-88
AT&T
  breakup of, 132-133
  worldwide affiliations of, 138
Automation
  in process innovations, 26
  in services sector, 125, 133, 137
Automobile industry
  complementary assets in, 71
  innovations in, 34, 38-39, 40
  standardization of, 74

H

I

O